IET MANAGEMENT OF TECHNOLOGY SERIES 25

Effective Team Leadership for Engineers

Other volumes in this series:

Effective Team Leadership for Engineers

Pat Wellington and Niall Foster

The Institution of Engineering and Technology

Published by The Institution of Engineering and Technology, London, United Kingdom

© 2009 The Institution of Engineering and Technology

First published 2009

The Institution of Engineering and Technology
Michael Faraday House
Six Hills Way, Stevenage
Herts SG1 2AY, United Kingdom

www.theiet.org

British Library Cataloguing in Publication Data
A catalogue record for this product is available from the British Library

ISBN 978-0-86341-954-6 (paperback)
ISBN 978-1-84919-100-5 (PDF)

Typeset in the UK by Techset Composition Ltd, Salisbury
Printed in the UK by Athenaeum Press Ltd, Gateshead, Tyne & Wear

For Ben and Matthew

Contents

Preface

The concept of leadership has been researched for generations.

Prior to the 1950s, theories which attempted to identify distinct, innate personality traits associated with successful leadership were prevalent. The goal of these studies was to find a 'single best way', in other words to define certain qualities or psychological traits that were common to all good leaders. The traits that were examined and considered to be of importance included intelligence, fluency of speech, physicality, and whether a person had an introvert or extravert personality.

In the next decade, researchers moved on from the concept of 'a single best way' to the assumption that different situations required different types of leadership – the behavioural approach.

Fast track to the 1980s, and a period of dramatic change when the old social contract of long-term employment in return for employee loyalty was ending. The model of leadership that emerged during this transitional period was called transformational, charismatic or visionary leadership. Margaret Thatcher and Richard Branson in the UK and Jack Welch of General Electric (GE) in the USA typify this style of leadership.

Contingency/situational leadership has subsequently emerged. Contingency theory says that no one leadership style is best in all situations. There is no single 'best' way. The team leaders' style is contingent or 'depends' on the situation. Success depends on variables: the nature of the task at hand, team member qualities, the readiness of team members to follow, etc. Situational team leadership, developed by Hersey and Blanchard,[1] is a type of contingency theory in which the team leaders' style varies according to the readiness of team members.

Recently, ethical leadership became a buzzword. The credit crunch has put paid to that and it has been replaced by regulatory forces as dominant influencing factors in the business world. Of course, if ethical leadership had delivered short-term results, the credit crunch and resulting drop in confidence might have been averted.

So where does this leave us now?

- How should we as newly-appointed team leaders develop a style of leadership that is right for us in a variety of situations?
- What are our key roles and responsibilities, and how should we perform these actions?
- What environment should we aim to create to make things happen?

- How should we manage and motivate others?
- If we have to handle interpersonal issues such as conflict, or to influence others to come around to our point of view, how should we do this?
- If we need to develop our own role, take a more strategic role and pass over part of our workload to others in the team, how should we go about doing this?
- And ultimately, how should we plan for our future in terms of career development once we move on from being a team leader?

All of these questions (and more) will be answered in this book.

Knowing that newly-appointed team leaders will have many pressures on their time and might not wish to read a book about team leadership from cover to cover, we have created this documentation as a Handbook, initially giving you key information about a particular topic, then bringing this information to life through case examples, checklists and, where appropriate, brief exercises to show in more detail how things work and what we mean by particular expressions.

What we have also aimed to do, is to *keep things simple*! We have not peppered the Handbook with management-speak or jargon, as we know from running team-leader workshops for engineers that this is the last thing you want. 'Suggest a range of best practice approaches we might adopt.' 'What do we need to be careful of?' 'Where have you seen this work well elsewhere?' This is the usual type of feedback we receive during training and development sessions. So we have borne this in mind while writing the Handbook.

So here we go. We hope you find this Handbook an informative, practical reference guide, useful in your day-to-day working activities.

Pat Wellington and Niall Foster
Europe Japan Management

About the authors

Pat Wellington is a Director of Europe Japan Management. She is an internationally renowned management consultant and author. Her areas of expertise encompass the management of change, leadership, customer care, interpersonal skills and business development. Her particular expertise within change management is how *Kaizen* (continuous improvement) can not only enrich peoples' lives at all levels within an organisation, but also bring very tangible results in terms of increased production and productivity. Her book *Kaizen Strategies for Customer Care* (Financial Times/ Prentice Hall, 1995) became a European bestseller when it was published, and she was also a contributor to *Kaizen Strategies for Improving Team Performance.* (Financial Times/Prentice Hall, 2000).

Pat has many years of practical experience as well as a thorough theoretical grounding. She started her working life as a general manager in a manufacturing and retailing organisation, and fifteen years ago moved into management training and consultancy. She has worked in various training and consultancy organisations, and recently headed up the Management Development Unit of London Metropolitan University, the largest educational establishment in the UK capital.

Her training and consultancy assignments have included working in blue chip organisations such as Coca Cola, Canon and Coats Viyella, and various United Nations aid agencies in Rome and Geneva. She has also worked in pharmaceutical organisations and with private hospital chains. In the Gulf she has worked with the Government of Dubai, and in Singapore and Malaysia she has delivered training programmes for engineers in various industry sectors. These have included a leading water authority, DSO National Laboratories, the Singapore Police Force and the Ministry of Education.

Niall Foster is a Director of Europe Japan Management. He is a pragmatic interim consultant with worldwide experience in human capital management, culture transformation, change management, start-ups, value based management, strategic planning, and competence development based on benchmarked best practices. Niall aligns an organisation's vision, strategies and employees with technology, processes and culture to realise bottom-line results. He diagnoses issues; he designs and implements/facilitates enduring measurable solutions for customers.

Niall's approach is at the cutting edge of development. His unique perspectives and methodologies are based on the assumption that organisations must find ways of tapping into the inventiveness of their staff and systems in line with their customers' changing expectations. Those he works with learn how to close the gap between concepts and action. He explains how staff attitudes combined with processes and systems can create environments where staff are better motivated, thus helping to bring about change and continuous improvement. All this is founded upon his expertise in human behaviour (training/coaching) and organisational design (consulting and selling).

Niall's work in Africa and Gulf has included Microsoft and the Arab Academy of Science and Technology in Egypt; the Great Man-Made River Project in Libya; Interlink College, Nigeria; Saudi Aramco; Al-Khafji Joint Operations and Marafiq Power & Utility Company for Jubail and Yanbu in Saudi Arabia; Qatar Gas; RasGas and Transfield Emdad Services/Intergulf; Kuwait Petroleum Corporation, Petrofac in the UAE; PDO in Oman. Since 2003 he has delivered public programmes and one-to-one coaching for Apex, EuroMaTech and Glomacs in Dubai.

While Niall is a consultant, he also leads considerable numbers of personnel in companies where he works as an interim director. He assists organisations that want to build competitive advantage through the development of employee talents and skills in support of their company vision and strategies. He was interim group training director with Barclays Bank for two years. Additional interim projects include Vodafone, the National Health Service, LA Partnership, Sterling Commerce, Henkel Chemicals, SAS Institute and Sandvik Steel.

Private sector experience includes Computer Associates, Cisco Systems, Nestlé, Dell, Toshiba, Unilever, Johnson & Johnson, Kraft Foods, State Street Bank, Coutts, Aviva Group, BT, European Bank of Reconstruction & Development, Cuisine de France, ABN AMRO and Hibernian Insurance. Public sector clients include the London Boroughs of Islington, Hackney, Lewisham and Kingston, the Ministry of Defence and London Transport. His non-executive board expertise includes SAS Institute, the LA Partnership and Stafforce, etc.

Europe Japan Management is a consultancy that specialises in the management of change, leadership, customer care and the performance improvement or *synergy* of teams. Continuous improvement (Kaizen) is incorporated into their consultancy and training programmes. They deliver tangible productivity and/or profitable results. Their expertise is founded upon their knowledge of human behaviour (training/coaching) and organisational design (consulting and selling). They align organisational objectives with team outputs to ensure measurable added value and a unique blend of the 'best of East and West'.

For more information search for Pat Wellington or Niall Foster on Wikipedia, or Tel: +44 (0)20 8467 8089
Email: pat@pwellington.plus.com or niallfoster@btinternet.com

Acknowledgements

This book is the product of many people's minds and efforts. Each has contributed insight, facts, experience, introductions and personal support during its preparation. Our contacts in companies around the world who have embraced change, have willingly given us time and access to information. This has helped us understand and record in these pages best practice approaches, and a step by step overview on how to handle all the day to day activities that team leaders and their teams need to undertake to deliver outstanding performance.

We wish to thank our friends and consultant colleagues who have shared with us their experiences, and contributed case studies used in the book. These include Chris Patrick, Bob Bryant, John Knapton, Mike Kearsley, Barbara Hawker, Professor Yochanan Altman, Jaroslaw Fotyga and Sheila Bates, all in the UK; Cynthia Elliott (Alexander) in St. Lucia, Adrian Tan in Singapore, Fouzi Al-Qassar in Kuwait, Mohamed A. Al-Temnah in Dubai, Amin Sadek in Tripoli, Eiman H. Mutairi and Peter Harlow in Dammam.

There are two colleagues whom we would particularly like to thank. First, Patrick Forsyth, who has consistently encouraged and supported Pat during the last 15 years of her career and second, Niall's great friend and colleague, David Buchanan, who gave us his time and invaluable access to Cisco Systems personnel in the preparation of this book.

Last, and by no means least, we thank Lisa Reading and her colleagues at the Institution of Engineering and Technology, who have advised and guided us on practical matters during the creation of this material.

Pat Wellington and Niall Foster

Part 1
Key principles of leadership

Chapter 1
Different types of teams

This chapter traces historically how people have moved from working in simple work groups to the creation of high performance teams in a business context in the twenty-first century. We will also examine the different types of teams that you might work in and the different styles of leadership that you should adopt according to particular situations. Finally, we will define the characteristics of a high-performing team.

1.1 Introduction

Over the years you have probably worked in a variety of teams. Sometimes they have been effective; sometimes they have not. A poorly led operational team, a cross-functional team that cannot establish a common objective, or a virtual team without robust communication methodologies in place can all end up floundering. Equally well, you might have worked in teams where there is energy, enthusiasm, and a great feeling of camaraderie. Performance standards are high; projects finish on time and on budget. Creativity and flair are actively encouraged, and 'blue skies' thinking results in a new range of services and products being offered to the marketplace.

Now all of the above examples did not happen by chance. There were a multitude of reasons why the teams that you worked in either flourished or floundered. So the purpose of this book is to give you as team leader the tools and techniques to create a team that flourishes and performs highly. This could be in a variety of teams where you take a leadership role. The duration of the team could be brief, say for a six-month project, or could be an operational team that is in place for years.

So what does it take for a team to flourish, and become a high performing team?

How does the style of leadership of the team have an impact on its performance?

And how did the concept of team working develop in the first place? Let's start from here.

1.2   History of teamworking

People have worked together in a supportive way since the dawn of history, hunting for food, growing crops and creating shelter, all in order to survive. Someone usually led while everyone else did what he or she was best at, and shared in the outcome.

People started to work in teams in order to overcome obstacles or to perform tasks which individuals could not hope to tackle. Teams have been brought together to build structures as diverse as the Great Pyramids and the Channel Tunnel, or to fight wars or harvest fields. Prior to the Industrial Revolution teams had an innate flexibility, so that armies of soldiers or labourers would form teams to fight or to build during the growing season, then disband and reform in their home villages for the harvest, repeating the cycle year after year.

Industrialisation and urbanisation destroyed this seasonal cycle, and management theories popular at the end of the last century broke down much of the 'cottage industry' teamwork into man-as-machine components of mass production. We can see how manufacturers are returning to work group teams, and how electronic communication enables close-knit and coherent teams to operate even though their members are geographically widely dispersed. New management theories such as de-layering, empowerment and corporate re-engineering, and the need for organisations to become more flexible and more specialised in the face of increased global competition and to retain their integrity during the current global recession, is leading to yet another incarnation of 'teams'.

Organisations, whether private or public, have woken up to the fact that they do not need to control production in order to supply a service or product, and that in many cases vertical integration is inefficient, since specialist providers can satisfy demands for 'components' more cheaply and to higher standards than in-house providers. 'Outsourcing', traditional for the input of materials and components to many industries, is now becoming common for services and intangible input as well as physical components, consumables and sub-assemblies. At the same time, electronic communications [notably electronic data interchange (EDI)] have facilitated 'just in time' (JIT) inventory techniques, and forced organisations to recognise that one key to competitiveness is long-term and close relationships with regular trading partners – teamwork, in fact. The supplier is no longer a commercial slave to be whipped but a commercial colleague to be nurtured.

So design and manufacturing teams are becoming more and more inter-enterprise, working alongside in-enterprise management teams.

The final step in this process is for organisations to 'deconstruct' themselves. Instead of being a monolithic structure, arranged in a pyramid with the chairman at the top, the workers at the bottom and multiple layers of fossilised management systems in between, the whole edifice is dissolving into small independent work groups, each able to 'sell' some service or value to others. The idea is for teams to come together, involving individuals or groups from around (and outside) the enterprise, perform some activity, be it on a project or long-term production basis, and then dissolve, with the components reforming into new teams to meet new challenges. Apart from a very small core of strategists and controllers the 'organisation' in the ordinary sense of the word is on the verge of collapse, to be replaced with something resembling a shoal of fish or flock of migrating birds, splitting, reforming and wheeling about to avoid dangers and to take advantage of currents and short-lived sources of food. With the aid of the electronic communications networks mentioned in the previous sections, many team members may work wherever they prefer to be, and the 'office' is becoming a pit-stop rather than a residence, with 'group-ware' replacing the conference room.

Meanwhile, as the framework dissolves, the teams live on, with key individuals perhaps being members of many teams simultaneously, providing personal and information links between them, and ensuring rapid flows of knowledge and 'best practice' from team to team. Teamwork has become the name of the game!

> 'Old ideas about size must be scuttled. "New big", which can be very big indeed, is "network big". That is, size measured by market power, say, is a function of the firm's extended family of fleeting and semi-permanent cohorts, not so much a matter of what it owns and directly controls.'
>
> Tom Peters[1]

1.3   What is a team?

In the current business world you will often meet a manager who will talk about his or her 'team', when in reality they are dealing with a group of individuals whose commonality of purpose is simply to prevent themselves being overwhelmed by the workload. Yet again, individuals working in a survival mode.

Jon Katzenbach, Director at McKinsey & Co., and Douglas Smith have been very influential in the study of teams. In their book *The Wisdom of Teams* they define teams as follows:[2]

- working towards a common goal
- the personal success of team members is dependent on others
- have an agreed and common approach
- the knowledge and skills of team members are complementary
- a small number of people, usually fewer than twenty

As Harvey Robbins and Michael Finley wrote in *Why Teams Don't Work*,[3] a team is about *people doing something together*. It could be a football team playing in a game against opposition, a research team developing a new pharmaceutical drug or a rescue team pulling people out of a burning building. The *something* that a team does isn't what makes it a team; the *together* part is.

1.4   Why work in teams?

'In unity there is strength' was a basic tenet of early trades unionists, who recognised that individual workers could be sacked or pushed around due to a visible imbalance of power between worker and employer, but if all the employees acted as one, the power was more evenly matched. The old parable, showing that a single twig can be broken by a child while a bundle of twigs is stronger than a grown man, is an alternative statement of the same concept. So we work as teams to pool our resources when addressing a challenge. In a trades union or a tug-of-war team this is very obvious, since everyone pulls together making, in effect, near-identical contributions and merging their individualities into a common effort.

In most business and sporting endeavours this is, of course, not the case. The team still works together to address a common goal, but we celebrate individual skills, all of which are needed to reach that goal. A soccer team comprises one specialist who guards the goalmouth, others who provide a more wide-ranging defence, then a midfield group with more general skills, and finally the forwards who are good at penetrating the opposing defence, and in setting up and exploiting scoring opportunities. Over the course of a season a well-balanced team will usually do better than ill-balanced teams, no matter how brilliant individual sections may be.

The same applies in business, where a successful team brings together people with a variety of specialist skills who each apply their particular talents to help the group overcome various obstacles encountered along the way. Some contributions may be fleeting, some will form the long-term core who do the bulk of the work, but all are necessary for a successful outcome. Let's look at a couple of examples.

For most of this century, assembly work has been done by specialists, each doing one task as the product takes shape. We have all see pictures of motor vehicle assembly lines, where the workers act as machines, fitting just a handful of components and then standing back to await the next vehicle and repeating the process. At best this is very efficient and effective, but it is an inherently fragile approach since if anyone falters the whole process is thrown into chaos. During the past decade or so a more robust approach has been adopted: team-working, where a group of workers together perform a major chunk of the overall assembly. Each worker soon develops a range of skills so they can cover for one another when an individual needs to pause, and individuals can move around, so relieving the tedium inherent in the 'man-as-machine' model; they can check each other's work with a positive impact on quality. Perhaps of even greater importance than these productivity-oriented benefits is the fact that the team grows into a social group with emotional support for individuals who are feeling low, and overall morale and motivation becomes much higher than that of socially isolated individuals on a traditional production line.

In designing and creating a new computer application a team probably revolves around the programmers, but will also involve specialists in the application subject matter, human interface design and perhaps database design. Hardware and communications specialists may be needed to specify the environment and provide inputs on suitable protocols and standard message formats; later on, documentation specialists, marketing, sales and training expertise will be needed to spearhead the distribution and implementation phases. Administrative and financial assistance will provide and schedule resources and monitor progress against timetable and budget and, if the product is for sale, legal advice may be required to handle intellectual property issues. This project-oriented teamwork, mainly involving individual contributors on a short-term basis, is obviously much more difficult to manage than the production-oriented teamwork described in the previous paragraph, and the social cohesion and resultant motivation has until recently been far more difficult to develop and maintain.

1.5   Research into teams

Numerous research assignments took place in the early 1900s looking, for example, at the behaviour of groups undertaking problem-solving activities, including the pioneering work of Professor Elton Mayo. In the period between 1927 and 1932 he conducted an experiment at the Western Electric Hawthorne Works in Chicago, which examined relationship between productivity and working conditions.[4] He looked at aspects such as humidity and brightness of lights. He also examined psychological aspects such as leadership style, group pressure, breaks and working hours. After extensive analysis the researchers agreed that the most significant factor that influenced performance was the building of a sense of group identity, a feeling of social support and cohesion that came with workers being far more involved with setting their own conditions for work, and their manager having a personal interest in each person's achievements. These research findings encouraged other organisations to consider grouping their employees into effective work teams.[5]

Moving on from the post-war period, extensive research was undertaken in the USA observing the behaviour of problem-solving teams, developing valuable work on group roles. Benne and Sheats[6] in the late 1940s probably pioneered this activity, but others, such as the highly successful social psychologists Krech and Crutchfield[7] and later C.J. Margerison[8] in the 1970s, also developed useful analyses of the processes which occurred in a business-based team environment.[9]

In the UK Meredith Belbin was also conducting research into teams at Henley Management College, and he framed nine roles characteristic of successful teams, concluding that if these roles are represented among team members and utilised in the team's operations then the team is predisposed to success.[10]

Belbin's nine team roles are well documented and there are psychometrics available to identify the personal characteristics of candidate team members so that, theoretically, it is possible to build a team whose members will collectively exhibit all the behaviours or roles required for success. However, the reality is that one rarely has the luxury of being able to build a team on this basis. If you use Belbin's roles as success criteria, then you have to ensure that they are delivered by training members to incorporate them in the processes of the team's operation, rather than by relying on their inherent predisposition to the chosen behaviour.

Belbin's early work also aimed to identify the qualities of a successful team leader. He sought a type, one in whom a set of specific characteristics and traits were to be found and which made that person a good leader. Starting from teams which were judged to be delivering a high level of performance, he backtracked to determine what it was the leaders were doing to facilitate the team's success. The concept of type proved elusive; he found that a wide range of behavioural patterns, competencies and strengths achieved success. He even concluded that superior intellectual ability did not necessarily make a successful leader, though if the leader could not keep up with the team, it predisposed both leader and team to failure.

More recently Belbin has provided us with a relatively simple set of guidelines for successful team leadership. While helpful, they are not, in my view, definitive. The matrix on page 14 summarises this thinking and is a useful first-level guide. I say first-level because in the hypercompetitive and fast-changing world in which we live, the team leadership role, in common with all levels and types of organizational leadership, is changing radically. We will explore some of these changes later in this chapter.

1.6   Different types of teams

A variety of teams can be found in an organisation:

- **Operational** – Teams you form part of by joining an organisation, such as HR, finance, marketing, etc.
- **Quality circles** – Teams throughout the organisation examining continuous improvement activities.
- **Self-managed teams** – Working together in their own way towards a common goal *which is defined outside the team*. For example, in a production line situation the senior management will have decided on the packaging/cardboard boxes to be used for containing a product, but the team will do their own work of scheduling/training/rewards and recognition.
- **Self-directed teams** – As above, working towards a common goal which *the team defines*, but also handles compensation and discipline and acts as a profit centre by defining its own future. For example, in a university a commercial unit could be set up to handle in-house assignments with corporate clients, and be reliant on generating sufficient income to cover its own costs and generate profit.

Some groups are formed for a limited period and for a clearly defined set of circumstances:

- **Task force/project teams** – Set up for problem-solving activities. This type of team will often be staffed by members of the organisation who have clout and who therefore are in positions of authority, able to implement the recommended changes that are required for problems to be resolved. Project teams are also set up for new product development and the creation of new services.

- **Cross-functional teams** – These are normally set up for a limited period in which to complete a sensitive task such as new product development or improvement to a process. They could be staffed, for example, by people from finance/marketing/HR and operations, and also could include a supplier or client. They need to draw on information from all parts of the business, including information from functional departments. System integration becomes important because it makes all information accessible through a single interface. They might need three types of information to be used for strategic, tactical and operational decisions, for example, with new product development. Cross-functional teams could also make decisions on inventory purchase or production scheduling and whether to test market a product or service.

How about the size of the team – does it really matter?

Ideally, it should consist of around eight to ten members, as over this number it is quite hard to handle all the different opinions, ideas and personalities; larger groups are inclined to break into smaller groups and then come back to pool information together.

1.7 Development of a team

In 1965 Bruce Tuckman described four main stages of group development that have been verified by research.[11] He explained that groups must experience various developmental stages before they become fully productive and function *as a group entity*. Some groups become stuck for varying lengths of time in a particular phase of growth. Thus moving through the phases of growth is unique to each individual group and learning happens at each stage. As Irvin Yalom says, 'a freely interactive group … will, in time, develop into a social microcosm of the participant members'.[12]

However, for optimum group performance all four stages need to be negotiated. These stages are **forming, storming, norming and performing**. In the counselling and psychotherapy world the fifth stage is vital; this is **mourning**.

When a group of people are put together to form a team, they don't necessarily know each other. There is unfamiliarity, and time and energy is spent assessing one another and jockeying for position. Focus on purpose and team objectives becomes secondary to these necessary preliminaries.

1.7.1 Forming (coming together)

This is like the first week on a training course – a gathering of disparate individuals trying to find out about one another. At this stage individuals are trying to assess attitudes and backgrounds of other members, as well as establishing their own personae in the group. Some members will test the tolerance of both the systems and the leader. On the whole, most members of the group are on their 'best behaviour' at this time.

Individual roles and responsibilities will be unclear, and there will be a high dependence on the team leader for guidance and direction.

1.7.2   Storming (open conflict)

Individuals now start to sort out and negotiate what they want from the group process (whatever the purpose of the group – learning or experiential). This is often a very uncomfortable period. Individual goals are stated and interpersonal clashes result when differences emerge between the goals/needs/wants of individuals. Alliances may be formed, sometimes even subgroups, and initial relationships can be disrupted. The main characteristic is conflict between group members. This is usually regarded as the essential stage of a truly representative group as it is, of course, a democratic process. Without the support of the team leader in negotiating this tricky stage, groups can remain secretly divided and less functional or creative.

If these differences are not negotiated or aired, individuals may operate within their own hidden agendas and hinder the healthy functioning of the group.

1.7.3   Norming (settling)

Agreement and consensus is largely formed amongst the team, and roles and responsibilities established. The 'rules' and 'norms' of the group have been accepted, such as no talking or interrupting while another team member is speaking during a meeting, how open communication is to be among the team, or practical considerations such as lunchtime cover. The team leader will delegate smaller decisions to individuals or a few people within the group.

1.7.4   Performing (executing task)

At this stage, members of the team are more strategically aware; they know clearly why they are doing what they are doing. This is the stage at which the group is functioning as a team. They are keen to get on with the task. Members take the initiative, cooperate and work interactively to meet stated and shared objectives.

As the team now has a good deal of autonomy, they can make decisions based on criteria established by you as the team leader. Obviously disagreements will still occur, but now they are resolved within the team. The team does not need to be instructed or assisted, and will come to the team leader occasionally for advice and guidance. This will enable the team leader to move on to key activities and take a more strategic role.

1.7.5   Mourning (letting go)

If a team is ultimately going to disband, this will be at the end of the task or project. Those people involved will no longer be meeting/functioning as a group entity with a common objective. Relationships between individuals may continue but this particular set-up will finish. This is often characterised by the stages of grief (which are not sequential – shock, denial, anger, bargaining, letting go/coming to terms and moving on, forever changed by the group experience and its loss).

Looking at this model, a team will often believe they are at the performing stage, when in reality they haven't moved out of the forming stage. To become a high-performing team they must go through all of these stages in the right order. The timing of the progression through the stages will obviously depend on the nature of the team. A functional team that has been together for years will usually take longer to go through these stages, and will almost certainly move backwards and forwards a few times as people leave the team and new arrivals have an impact on the group dynamic. Project teams, on the other hand, who are just together for a short period of time, say six months, will have to develop and go through these stages at a quicker pace to be truly effective. Teams of any type can easily get stuck at one stage, say the storming stage for example, if roles and responsibilities do not appear to be fairly distributed, or if the team cannot agree a common goal or way forward.

1.8   A major step forward in the concept of team development

While there was extensive research underway in the USA and UK, as we have indicated above, what really made the difference to the concept of team development was the impetus from Japan following the end of World War II. A group from the USA including W. Edward Deming and Joseph Juran went to Japan to help rebuild industry there. Deming has become known as a 'prophet unheard' because his philosophy and approach to business was ignored in his own country. When the team initially arrived in Japan, product quality was so poor by international standards that cheap goods such as radios were being produced where the knobs fell off! The Japanese were impressed by the sheer profusion of goods being produced in America, and were more than happy to listen to any American experts who were prepared to come to Japan to talk about quality.

Deming and Juran introduced a new management style to the Japanese based on the premise that staff at all levels should become skilled at more than one job or activity, and importantly should be responsible for quality. If someone on the production line at Toyota saw that quality was not up to an agreed standard they could pull a cord placed just above their head, and stop the production line until quality had been restored. This was a radical divergence from manufacturing in the West. In the manufacture of cars, for example, each person on the production line would be responsible for just one activity, and quality would be assessed at the end of the production line by a quality department.

Japan had also become very short of natural resources following the war, and so the concept of JIT inventory control was introduced, with raw materials being delivered when they were required rather than being stockpiled. The reduction of waste – *muda* – was seen as a vital activity in those challenging times. This whole new system of production was dubbed the Toyota Manufacturing System by the car manufacturer itself, and 'lean production' by almost everyone else.

The fundamental difference introduced by Deming and Juran was the new style of management that underpinned team working. This management style was based on *Kaizen*, continuous improvement, which proposed that the culture and values of the organisation needed to support staff, to nurture their development and to challenge people to think of new approaches to past methods. Enablement/empowerment of staff was developed and actively encouraged. Customer focus became key in this approach, together with working far more closely with suppliers (the *keiretsu* system) than organisations had done in the past. Historically, Western organisations had selected suppliers based on tying them down to low-cost supply, whereas in Japan suppliers were being involved in a fairer long-term arrangement.

By the early 1980s Japanese firms were beating Western organisations in everything from price to quality, and plane-loads of managers were visiting Sony and Toyota to find out about lean manufacturing, the consensus form of management and the building of teams. Total quality management in America was the most influential activity in this period.

If we cut to the present day, this style of management is the basis for how organisations throughout the world manage staff and develop team work. Peter Senge's book, *Fifth Discipline*, published in the early 1990s, has facilitated the development of the learning organisation.[13] Continuous improvement is adopted in every industry sector, and adapted according to local conditions and requirements. Lean manufacturing has been transported via joint ventures with the Japanese to other parts of the world, and Six Sigma has emerged from the Motorola manufacturing system. The principles of working more closely with suppliers is now a given with organisations such as Marks & Spencer, Motorola and Chrysler. The Japanese have themselves had to learn from the West, for example in the area of *keiretsu* where this is no longer the model of best practice. Toyota have had to turn to Ford to discover how to improve the relations between engineers and its shop-floor workers, and to Chrysler to learn about 'value engineering', a new way to speed up car production by using interchangeable components in different models.[14]

Synergy and adaption is the name of the game!

1.9   Characteristics of high performing teams

'It's surprising what a team can achieve if everyone forgets about his personal gain!'

Blenton Collier, American National Football League Coach

The first condition for good teamwork is that each member of the team is aware that *he/she alone* is responsible for the whole. It is not shared responsibility, it is the responsibility for everything on the shoulders of each one.

The goal is to realise the **HOLOGRAPHIC TEAM**. If you take a holographic image of a house, for example, and cut this image into pieces, you find the image of the house on each piece.

The state of mind of each team member reflects the state of mind of the whole: it is the definition of **TEAM SPIRIT**.

What are the necessary conditions for reaching the ultimate stage of the holographic team? What should it share?

- certain values (at least the three basic values)
- certain skills
- certain foci of interest
- fun
- confidence
- and ... **A GOAL!**

1.10   Areas of team focus

There are five areas of focus which all teams must address in order to function effectively:

1. **The organisational environment**
   Are things outside the team helping or hindering performance?
   Does the team have adequate resources (people and financial)?
   What about policies, strategies, structures, the market?
   Is the team's mandate understood and widely accepted in the organisation?

2. **Goals**
   Do people understand and accept the team's primary task (its
      organisational mandate)?
   Are team members involved in setting objectives?
   Do all individuals agree with priorities?
   Is progress towards team goals regularly reviewed by the team?

3. **Roles**
   What do team members expect of one another?
   Are these expectations clear? Acceptable?
   How are conflicts in expectations handled?
   Is unnecessary duplication avoided?
   Who does what?
   Do team members know how their personal efforts contribute to the team's
      success?

4. **Processes**
   How is information flow and the need for coordination handled?
   Are all team members coping with current technologies?
   How are problems solved, decisions made and adhered to?
   Are meetings efficient and improvement orientated?
   Are deadlines and milestones clearly established and agreed to by the team?
   Are reporting procedures clear and concise?

5. **Relationships**
    How do members treat and feel about each other?
    Are members' needs for recognition, support and respect adequately met?
    Is there effective analysis and feedback of group and individual
        performance?
    Are good communication channels established with other teams, and parts
        of the organisation?

1.11   Chapter summary

- Teams have been in existence since the dawn of time, and have acted as a source of help and support, and aided in performance enhancement.
- Research over the years has highlighted the needs and requirements of staff and what it takes to work effectively as a team.
- Deming and Juran's arrival in Japan and the management style they introduced revolutionised the way staff were managed and has ultimately had a profound effect to this day.
- This style of management and the implementation of teams has subsequently been adopted and adapted by companies around the world.

Chapter 2
Role and responsibilities of team leaders

This chapter provides an understanding of the role and responsibilities of team leaders. It gives the team leader guidance on those things you will need to consider and get right – how to control change, reduce risks, protect productivity and retain the team spirit necessary to achieve high performance and realise your team's objectives quickly and decisively.

2.1 Introduction

Being a team leader is an incredibly rewarding and valuable experience. It's tough. It's demanding. It is also an opportunity for you to make a positive impact in so many ways. The benefits of being a team leader are many and include:

- **Skill development** – A chance to utilise and develop key workplace and management skills.
- **Career development** – Leading teams and projects are experiences that enhance your CV.
- **Satisfaction** – You can feel a genuine sense of achievement. How might it feel to be able to make things happen?
- **Improved network** – It's a great way to meet new people inside your organisation and to enhance your career prospects.

Your job, your role, is to manage. As a team leader you take responsibility for making each working day happen. Make things happen *with ease*. Help your team with the process and with the 'how' decisions. Your job is to maintain a safe working environment while achieving your objectives and making day-to-day challenges an enjoyable experience for all of your team. A balance must be maintained between keeping your team focused on the task and allowing them to make their own decisions (Figure 2.1). You should never 'pull rank' or apply undue influence on or 'command' your team.

In order to get the most out of being a team leader and to deliver maximum impact, it is crucial that you know how to *plan* effectively, *implement* your plan efficiently, and *review* your experience to ensure mistakes do not recur. There should be tangible

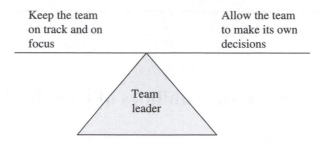

Figure 2.1   Maintaining balance between objectives and day-to-day decisions

benefits for internal and external stakeholders. You should build partnerships that aid the team's work and guarantee you receive the resources you need for success.

2.2   What is team leadership for?

What is team leadership for? This may seem a strange question to ask. 'Managers and team leaders are there to run the company, and their teams, of course', is a typical response. But what does that mean, and to whom does it apply? Let's begin by thinking about team leaders in terms of how they spend their time: in internal meetings, reading or writing reports and supporting, training and performance-managing members of their team. Anthropologists would probably coin a word to describe all of these essentially 'in-organisation activities', and might call it 'administration'. Then they might observe a second set of activities concerned with 'outside team activities' – meetings with suppliers, managing projects, attending conferences and so on – and bundle these under 'representation'. So what team leaders do can be summarised as administration and representation, but this doesn't really answer our question.

'Leading' implies two things: a group of followers, the employees and goals or objectives to define direction. Some objectives are fairly trivial: evolved products and services, higher productivity, reduced costs and so on. Taken together, though, they have a 'higher' purpose, the survival of the organisation, since upon this depends the survival of its components, the health of at least some of their trading partners, the continued employment of their staff and the regular appearance of everyone's pay cheques.

How do they go about this? Simple observation shows that good leaders make their organisations work *effectively* and that they prepare and *execute* plans to move their organisations forward. Bill Gates, Jack Welch, John D. Rockefeller, Richard Branson and Catherine the Great are just a few managers who are known (or remembered) for effective organisation, preparation and execution of plans. However, these alone are not sufficient. After all, Napoleon Bonaparte and Alexander the Great certainly oversaw effective organisations, but their 'organisations' died with them. At the simplest possible level this is because they did not plan for the future and construct means for their empires to continue after their founders' departures.

So we may conclude that what team leaders are for has to do with looking around the world in which their organisations exist, setting objectives to avoid potential hazards and capitalising on opportunities, then leading their team toward these objectives.

2.3   Today and tomorrow

'No level of efficiency would have kept a stagecoach builder in business today.'

Peter Drucker [1990]

The management guru Peter Drucker makes a valid point above. New technologies have rendered the stagecoach obsolete, and suppliers of thousands of other products and services have seen their markets wiped out by better solutions to users' underlying needs. So it is at the core of the task of the team leader to ensure that the organisation's products today are not tomorrow's stagecoaches.

Efficiency means doing things in such a manner that the customer is kept happy (and will come back for more), the members of the team are happy (and will wish to remain in your team), the shareholders are happy (and will continue to give you financial support), and presumably you will be minimising waste as well. Good, effective team leaders run efficient 'shops', but here you are concerned with the longer term. Peter Drucker again: 'efficiency is only a matter of doing things well; effective power is doing things advantageously' or, in simpler words 'Don't work harder, work smarter.'

If team leaders' first responsibility is to ensure the survival of their companies, thereby ensuring a continuing flow of benefits to all concerned, then the objectives they set must acknowledge what is happening in the outside world. And there we see change.

Team leaders have to react to changes in the operating environment: new competitors, new technology, new legislation, and new ways of exercising 'effective power' and gaining maximum leverage from the human, financial and material resources available. If we look back over companies which have failed (or been swallowed up by predators) during the past few years, it will be clear that many failures have resulted directly from an unwillingness or inability to react to change; others may have failed by reacting inappropriately. Successful companies, on the other hand, not only react to external changes, but they often initiate change and so wrong-foot competitors who have to change direction in order to catch up.

It follows that a successful team leader has to demonstrate a willingness to decide and help shape the organisation's future, recognising and taking risks, seizing opportunities yet avoiding future dangers. Here we complete our definition of what team leaders are for: **to lead their teams along a safe path through an uncertain future**. In other words: the manager is the agent of change. This applies to all team leaders and managers – the individual who runs the post-room, who looks after product development or marketing, or who sits nearer the top of the pyramid and exercises wider responsibilities. They all have the same task, to make it work efficiently today and initiate and oversee those changes which will make the organisation fit for tomorrow.

2.4   The team leader's role

Typically, team leaders are first-line managers critical to organisational and team success. Just look at a typical financial institution job description below and compare this with the team leader's role as described in Figure 2.2.

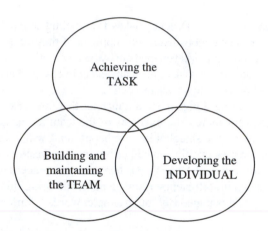

Figure 2.2   The team leader's role

Title: Support Team leader	Level:	Code:
Business line:		

Background

The Support Team is accountable for ensuring that the department works in partnership with its business and internal customers and is successful in meeting their needs as outlined in the customer guide. The Support Team will work in partnership with any area that is impacted by the department to provide speedy solutions to issues that impact delivery, design, planning and any process or improvement deemed necessary to ensure business and internal customer requirements are met. The Support Team is responsible for effectively supplying opportunities to improve business performance and to enhance the image of the department by its ability to deliver business objectives.

Purpose of the role

- Responsibility for managing all Support Team projects as agreed with your line manager.
- Direct responsibility for all day-to-day operations and processing activities of the Support Team. Particular emphasis on 'customer' interface, team members' performance and quality of service rendered to business and internal department customers and management.
- Be the principal person responsible for preparing individual project objectives, plans and specific outcomes and for defining all internal and external standards.
- Ensure all team members comply with agreed standards of service, control and performance.
- Plan, prepare and monitor the Support Team budget.
- Coach the Support Team members and ensure they possess required skills and a 'solution thinking' mindset.
- Follow up and monitor progress of completed projects within four weeks of sign-off to ensure implementation remains on track.

Key responsibilities and approximate time split

Identify, recruit and retain team members 20%

- Lead work jointly with Fraser Murray to agree strategy for Support Team recruitment:
    - determine Support Team recruitment needs.
    - produce role profiles.
    - interview and select team members at assessment centres.
    - manage team members' performance development.
    - design and run pre-joining events for team members and provide direction for the definition of a Support Team induction programme.

Project management 50%

- Personal responsibility for ensuring the Support Team meets the needs of each business project:
    - ability to negotiate for additional support and assistance to ensure success is a prerequisite.
    - ensure the team structure remains relevant; review projects regularly and make team changes where necessary.
    - conduct benchmarking exercises after each project and implement improvements where appropriate.

- Formulate practical and measurable solutions via the Support Team members for problems/issues to ensure that the department's solutions are aligned within all functional areas and satisfy the needs of the business.
- Provide weekly status reports to line management/others as appropriate that recommend strategic solutions to problems/issues.
- Maintain effective business relationships with key stakeholders.
- Identify improvements to L&D processes and practices and support implementation where appropriate.
- Communicate with all departments to enable the formulation of policies/practices/processes that anticipate changes in legislation, regulation and business needs, which must be reflected in all project solutions.
- Proactive communications, liaison and evaluation with internal and external customers.

Team communication, management and monitoring 30%

- Raise the profile of the Support Team wherever and whenever possible.
- Provide clear direction for team members.
- Provide support and guidance to team members.
- Undertake regular feedback and coaching sessions with team, providing necessary support for individuals' self-development and performance improvement.
- Determine and manage events required for team members to increase their knowledge and language of business customers and their objectives.

Key personal attributes

- developing/coaching people
- managing relationships
- influencing others
- integrity
- decision making
- credibility
- supporting personal attributes
- meeting customer needs
- business awareness
- communication

Skills required to undertake the role

- interpersonal skills
- presentation skills
- influencing and negotiating skills
- leadership skills

- time management skills
- coaching/counselling
- ability to continually identify and implement business improvements

Knowledge of the organisation's products, services and policies and/or other specialist knowledge required to undertake the role

A good knowledge and understanding of the department and the business direction and structure of its customers.

Additional details of exceptional aspects of the demands of the role

This is a high-profile demanding role with multiple responsibilities. An ability to work effectively under pressure and prioritise own and team's workload is essential as is attention to detail and a willingness to work extended hours in a particularly hectic and busy environment.

2.4.1 Achieving the task

Achieving the task is your primary purpose. You *must* do this. No excuses! We've come across team leaders who seem to do everything but what they should do. Focus on concrete (what exactly and by when exactly) results. Go for measurable gains. You will agree your team's objectives with your manager. Make sure that whatever you agree leads to the most urgent operational improvements. Focus on those things that go straight to the bottom line or that contribute directly to your competitive position. This is how to get noticed. The idea here is that there must be measurable benefits to stakeholders based on the team's activities.

To get results you must move with authority, make decisions, act. You must consciously influence and use your power as a team leader. You can't mobilise your team members if you're immobilised yourself. You won't achieve anything without 'power' and if you can't do anything, why are you in charge? Team members will not follow someone they don't believe in and they won't believe in you unless you believe in yourself.

If your manager cannot do anything to help you is s/he actually a manager? No! The same applies to you as team leader.

2.4.2 Building and maintaining the team

Building and maintaining the team is critical. Ensure each team member knows what's expected of him or her. Don't leave anyone to figure things out on their own. Get rid of ambiguity. Nail down every team member's responsibilities with clarity, precision and attention to detail.

This means that there must be no grey areas about where one job stops and the next one starts. Clarify the responsibilities each team member is supposed to shoulder.

Figure out precisely what needs to be done, who's going to do which part of it, and communicate your plan. Give every team member a brief job description. State your expectations regarding standards of performance. Describe the chain of command in the team. Outline each person's spending limits, decision-making authority, and reporting requirements. Everyone will be best served if you put this information down in writing.

Check to make sure that each team member understands the team's (whole) set-up and how it fits together. Be careful to avoid job overlap, since that feeds power struggles, wastes resources and frustrates everyone involved. When explaining to team members what to do, also specify what they should *not* do. Differentiate between crucial tasks and peripheral, low-priority activities. Spell out what needs to be accomplished in each position, by when, and which team member will be held most accountable. Once you have done this, pay attention to what team members are doing. Keep everyone on track. If you see something going wrong, fix it immediately.

Of course, team members can help design the team's priorities and objectives. You must consider their input. The more they can shape the agenda, the more buy-in and commitment they'll show. Plus, their ideas might dramatically improve your sense of priorities. In the final analysis, though, *you* remain accountable. The buck stops with you.

Team members need to obtain a clear sense of direction quickly. How else can they be effective as a team? Clear priorities help team members figure out how to spend their time. Set short-term goals that the team can achieve quickly. Make sure your instructions are unequivocal and easily understood. Make known your commitment to them. Get them to commit to achieving their goals. Tell them from the start to expect some mid-course corrections. The agenda will have to adapt as the situation demands it. But, always keep it clear and communicate it constantly.

2.4.3   Developing the individual

If you have the right team members to begin with you won't be forced to make changes later on. When you become a team leader, get to know your team. Approach the exercise as if all team members were 'new hires'. Check for people's adaptability. Ask yourself who is best suited for which role. If there are weak players, position them where they'll hurt the team least. Size up your team with a dispassionate, discerning eye. You need good data, and you need it in a hurry. You can't afford to sit back and figure out your team members as the months go by. You need to make informed judgements *now*. If you don't trust your skills at this, or if you feel that you can't make the time, get help.

Look for strengths, weaker points, aspirations and work preferences, experience and areas of expertise, concerns and points of resistance. The sharper your insights into each individual, the better the odds that you'll manage him or her effectively. You must lead, but also be able to encourage and motivate, educate and train, if necessary to nurse, and to trust the people whom you manage. Above all, as the team leader you have to make decisions. Since arbitrary decisions lead to chaos, they must be based upon information, and, of course, no team can embark on the implementation of a decision until they have been given this information. Most projects require sanction from other departments, upper managers and sometimes external agencies, and their decisions also involve information.

Your key people can be the cornerstones of your team effort, so don't take these people for granted. Make everyone feel important. Invest the same time and effort in creating the team as you would in recruiting a new employee. Enthuse them about the work in hand. Ensure your team members are on board emotionally.

2.5   The team leader as an agent of change

You have now seen that team leaders have to recognise and seize opportunities for change, identify the strengths and weaknesses of the organisation, capitalise on the former and overcome the latter. They have to provide leadership in part by making judgements about the unknowable. However, at the day-to-day level they have to maintain stability of projects, sales, production and distribution systems with their allied financial systems, in order to avoid chaos. We know that every system may be improved, but incremental changes must be accommodated in a secure and stable structure, preferably in a pleasant, safe and friendly environment. Creating and maintaining such an environment is an essential task: you might call it the 'nurturing' side of management.

Team leadership means management. Management is the effective use of resources to achieve your company's and your team's objectives, and the primary objective is the survival of the company.

In a competitive world, with rapidly changing technology and expectations, the company has to evolve or die. In the short term and under the guidance of its managers, it must react to regulatory and market stimuli and protect itself against accident. It must also look to improve efficiency through a programme of continual incremental change of processes and systems, and must continually enhance the product line (and level of customer service) to meet customer requirements.

Medium-term changes can include termination of unnecessary activities, taking existing products into new markets or territories and refocusing products for new market sectors.

Long-term changes combat the 'stagecoach syndrome'. Options include major rethinks of product lines to incorporate new technology, keeping abreast of changing customer requirements, and transferring resources to new enterprises.

The team leader's role is to plan the changes and to coordinate their implementation. This involves both administering and nurturing on-going activity and leading the team through change. Modern business managers gain their authority by demonstrating proven success and good interpersonal skills. Formal and informal information networks are essential prerequisites for defining objectives and making plans. Information exchange is also an essential element of morale building, and without high morale the team will not retain high-quality staff.

And without the staff, the company is nothing.

> 'Change masters are – literally – the right people in the right place at the right time. The *right people* are the ones with the ideas that move beyond the organisation's established practice, ideas that can form into visions. The *right places* are the integrative environments that support innovation, encourage the building of coalitions and teams to support and implement visions. The *right times* are those moments in the flow of organisational history when it is possible to reconstruct reality on the basis of accumulated innovations to shape a more productive and successful future.'
>
> Rosabeth Moss Kanter, *Change Masters* (1983)

Case Study: Middle East Hospital

We recently agreed a change project with a Saudi Arabian Health Care provider who in 2008 had a patient base of 350,000.

The project objective was to help the organisation to implement its vision, of being the best medical provider to it's customers. We began in two major departments, the Emergency Room and Primary Care.

This project aim was to establish in the mind of all staff best practice customer care, behavioural, standards. Additional project aims included:

- inspiring all staff to deliver a customer care vision that created shared meaning and understanding
- setting organisational direction and establishing customer care performance improvement strategies in unit/individual action plans
- shaping and strengthening the competence of staff to create long-term sustainability

How was this realised?

This was done by defining a new *Corporate Standard* for customer care. This new Standard was designed with, and by, staff. It clarifies to staff the customer care service the organisation aims to provide its patients and their dependents.

> A customer care toolkit was also created. The toolkit served as a model to drive consistent behaviours to customer care initiatives. The toolkit was a concrete guide that explained to staff how the Standard should be delivered.

To embed the standard as part of a culture change process, policies and procedures needed to be impacted. This was done by addressing the following policies and procedures as detailed in the diagram below.

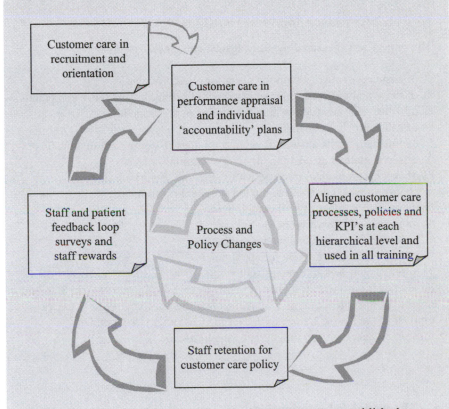

A customer care committee made up of senior managers was established to recommend decisions. Terms of reference were prepared for this committee as working guidelines.

Customer care behaviours were clarified at each hierarchical level. This detailed what customer care 'is' and 'is not'. These behaviours were then aligned with the company's appraisal system, job descriptions, orientation programme and departmental goals. Customer care as a competence was added into their appraisal process.

A Service Level Agreement was prepared to identify *internal* responsibilities and behaviours of staff. This defined acceptable levels of service between staff and the methods used to monitor and report performance that benefited staff and customers alike.

A project communications plan was prepared to inform managers and staff of the Standard and explain what they must do to align individual, team and departmental performance with it. A newly devised staff reward scheme included a customer care element.

Future project elements will include transforming the toolkit into an eLearning module using a modern interactive format.

How was this change project measured?

The project was measured monthly against an agreed baseline:

% of customers satisfied with the overall service provided
% of those making complaints satisfied with the handling of those complaints
% of office calls answered within three rings or 15 seconds
% of call centre calls abandoned
% of call centre calls answered within 5 minutes of being held in a queue
% of score attained against the 'mystery shopping' criteria
% of cases (complaints) resolved within 2, 3 and 7 working days
% change in each clinic of the top ten actionable complaints
% increase in number of staff compliments
% improvement in staff morale
% improvement in visitor survey response
% improvement in wait time for appointments
% improvement in wait time in facility
number of front-line staff trained in implementing the Customer Care Standard
number of managers trained in implementing the Customer Care Standard
number of facilitators trained to deliver the Customer Care Standard
roll-out of customer care programme kingdom-wide

2.6   The leader-led relationship

Corporate success centres on the relationships between team leaders and their teams, which in turn reflect the subjective interaction of two personalities. Common team-leader traits which tend towards positive team relationships include:

- enthusiasm about the organisation, its products and objective
- honesty and openness in communications, to build up trust
- technical ability, plus questioning skills to help staff solve problems
- decisiveness and the courage to defend decisions and, if need be, to admit error
- sensitivity to people's feelings to avoid causing needless hurt and open the way to small kindnesses

Team leaders who exhibit these traits gain good reputations which will filter out from their immediate circle; conversely, poor judgement in these areas will be simplified, exaggerated and passed around the organisation.

The purpose of management/team leadership is to enable the organisation/team to attain its goals. The organisation or team is just a collection of people working in concert: getting the best out of your people is the way the team leader achieves his/her purpose.

2.7   Principles of *human* team leadership

In the mid-1980s Ken Blanchard wrote a highly successful book called *The One Minute Manager.*[1] It epitomised the goal of a simple and easy answer to management problems and tasks. We suspect this may have been the final toll of the bell for faddish and simplistic approaches to the complex and challenging task of management. Indeed in 1991 the *Harvard Business Review* headed a marketing mailshot with the exclamation, 'The One Minute What? We think we are being insulted ... management is a much more complex business. Who heard of The One Minute Surgeon?'

However, team leadership is just that: leading teams of people. You must never forget the human side of managing. After all, your success comes from your team members. There are four key principles of human management that we recommend you never forget.

1. **Set an example** by demanding the most of yourself.
2. **Demand the most of your team members** to give them opportunities to be proud of themselves!
3. **Personally provide your team with as much assistance as possible.** Why do they need you?
4. **Protect team members from fear** and help them to overcome it.

2.8   The team leader's responsibilities

Your responsibility is to minimise risks, protect productivity and enthuse your team members so they perform well and realise the team's objectives decisively. What follows is a practical guide and review of the advice necessary for getting team roles and responsibilities 'right', to coordinate team effort and to set direction, targets, goals and purpose for the team. Understanding these will help you monitor your own performance and that of your team.

What is expected of you? Your responsibilities are varied and can include the following:

*   Focus team members upon the purpose and tasks of the team and projects; remind the team of the outcomes the organisation, team itself and service users seek and how these fit with the organisation's vision, mission and goals.
*   Perform effectively in clearly defined functions and roles.
*   Promote values for the team and demonstrate the values of good team leadership through your own behaviour.
*   Take informed, transparent decisions and manage risk.
*   Develop the capacity and capability of the team.

- Engage stakeholders and make accountability real.
- Organise the team, securing resources (space, time, people, budget) and clearing organisational boundaries.
- Work closely with others to create agendas for (team project) outcomes and ensure next steps are assigned.
- Assist others or facilitate meetings; encourage quiet team members to speak up, and when necessary shut down those team members who tend to dominate.
- Recognise and celebrate accomplishments.
- Communicate with others in the organisation regarding the team's progress.
- Work safely and ethically.

In today's business environment, coordinated team effort is required to achieve business objectives. It isn't easy, increasing productivity and adding value/profitability to organisations which constantly have to reinvent themselves due to the pace of change. Everybody gets stressed, distracted and confused. Some jockey for position. Some give up. Team members look at *you, the team leader*, to 'fix things', and even those who are willing to help can't agree on *how* things should be fixed. Senior management is telling you to do more with less. 'Do it better', they say, 'it's the only way the organisation can survive in today's competitive marketplace'.

2.9   What is the team leader's specific responsibility?

Given the long list above, what is the team leader's specific responsibility? How do you answer this question? Here are some typical answers: 'realise the team's objectives', 'motivate staff', 'ensure you have all the resources you need', etc. Of course these are correct. However, the specific responsibility of any manager or team leader is to

CONTROL CHANGE!

This does not mean *reacting* to change! You must be proactive. You must spend more time changing and planning or preparing for the future rather than reacting to past or present mistakes.

PROACTIVE

The telephone rings and you answer or react to the ring tone. How much of your time is typically spent *reacting* to situations at work? If you react more than 50 per cent of the time you will never be able to control change. How often do the same problems or mistakes recur in your workplace? If this happens a lot it says one thing: some manager is not doing his or her job of controlling change, and so the whole organisation suffers. This is caused by more senior management *not* doing their job properly and controlling change. They are not proactive.

What is the most significant change facing you personally in your organization today? Tom Peters said, 'a profitable company continually adapts, responds rapidly to change and creates change in accordance with the wishes of its customers'.

How can the team leader control change on a day-to-day basis? You do this by obtaining information, asking questions, listening and establishing concrete objectives.

Focus the team on the overall goal or mission of the team and provide resources to help the team reach that goal. The team leader keeps the team on track. Do not think that you are 'in charge'. Sometimes you need to lead and sometimes to facilitate. You should not place undue influence on the team and its decision making.

2.10   Team leader or team facilitator? You decide

Sometimes you need to be the team leader and sometimes you need to facilitate. To develop team members the role of facilitator may be rotated from team member to team member. It becomes clear then that *the team leader's role is to move the team through the process. Facilitators should avoid making decisions or offering opinions.* The facilitator never decides; he or she simply ensures that the required result is understood and that the decision belongs with those responsible.

Below are some responsibilities and quotes. Indicate whether you think the responsibility or quote is attributable to either a team leader or team facilitator by selecting the appropriate answer.

Exercise

		Leader	Facilitator
1.	Ensures team members know the purpose of the team and its overall goal.	☐	☐
2.	Reviews the agenda and gets input from all team members regarding its appropriateness.	☐	☐
3.	Secures meeting rooms, flip chart and other materials. Clears organisational barriers that may impact the team.	☐	☐
4.	'You've completed your agenda. Is this a good time to set the next meeting, or does the team want to review the decisions made so far?'	☐	☐
5.	Recognises and celebrates accomplishments.	☐	☐
6.	'I think posting progress reports is a good idea. However, that's not what we're here to discuss. Remember, our only purpose is to select a supplier'.	☐	☐

7.	'You wanted to select a supplier. You're now talking about progress reports. Does the team want to continue discussing progress reports and table the supplier selection? Or do you want to return to the discussion about suppliers? You only have 10 more minutes available'.	☐	☐

Feedback – you decide

Compare your responses with the feedback given below.

1. **Ensures team members know the purpose of the team and the overall goal.**
   The correct response is leader. The leader is first and foremost responsible for ensuring everyone knows the purpose and goal of the team.

2. **Reviews the agenda and gets input from all team members regarding its appropriateness.**
   The correct response is facilitator. Facilitators should begin meetings by reviewing the agenda and ensuring everyone understands the process.

3. **Secures meeting rooms, flipcharts and other materials. Clears organisational barriers that may impact the team.**
   The correct response is leader. Leaders provide meeting materials and should work to make sure the organisational barriers are removed, to allow the team to achieve its goals.

4. **'You've completed your agenda. Is this a good time to set the next meeting, or does the team want to review the decisions made so far?'**
   The correct response is facilitator. While a leader might say this, the most likely person to remind the team of their progress and ask for input is the facilitator.

5. **Recognises and celebrates accomplishments.**
   The correct response is leader. Leaders can recognise team accomplishments with small celebrations or special events. It might simply be snacks or humorous certificates. Regardless of the item, teams enjoy being recognised.

6. **'I think posting progress reports is a good idea. However, that's not what we're here to discuss. Remember, our only purpose is to select a supplier'.**
   The correct response is leader. Sounds like someone reminding the team of their purpose, doesn't it? And the person began by sharing an opinion. That's something a facilitator would never do.

7. **'You wanted to select a supplier. You're now talking about progress reports. Does the team want to continue discussing progress reports and table the supplier selection? Or do you want to return to the discussion about suppliers? You only have 10 more minutes available'.**

The correct response is facilitator. Same comment on a team issue, but this time by the facilitator. Note that the facilitator keeps personal opinion out of the comment, and the facilitator poses a lot of questions. The team is in charge, the facilitator simply keeps them moving.

2.11   Team leader and team facilitator – roles and responsibilities

Makes things happen with ease

Team leaders develop and manage the process for the team, helping them stay on track and proceed through ground rules ('You agreed everyone should participate, and we haven't heard from Bill. Bill, would you like to share your reactions to the discussion so far?')

Helps the group with the process

Have you ever been a member of a team where only two or three people made all the decisions? That's not a good process. Facilitators help the team through situations similar to that by neutralising potentially dominating people. The result is a process reflecting the group's collective brain power, not simply the ideas of two or three people!

Helps the team with the 'how' decisions

Elsewhere in this book are descriptions of several tools and techniques team leaders can use to help make decisions, such as brainstorming, and prioritising. Team leaders are skilled at using these tools and helping teams realise their potential.

Sometimes you need to be the team leader and sometimes the team facilitator. The team leader keeps the process moving forward by making suggestions, never dictating (unless all else fails and in the case of emergencies where snap decisions are required).

Read the table below for some suggestions as to how a good team leader might keep a team on track.

Task/situation	What the team leader might do/say
Review agenda	'Does this agenda look like a good road map for today? If the team accomplishes this, will everyone agree that your goal was accomplished?'
Establish ground rules	Review possible ground rules and decision-making processes. Require that everyone speak up on the ground rules.
Diverge from the topic and agenda	'You agreed you wanted to make decisions about "X". You've just spent 10 minutes discussing "Y". Do you want to *continue* discussing "Y" *or table discussion for another meeting and return to discussing* "X". You haven't made a decision about "X"'.

Note that the team leader *never tells the team what to do*. Rather, you should remind the team of its agreed-upon ground rules and prompt them to consider their options. In the last example, maybe the team really wanted to abandon 'X' and begin a discussion of 'Y'. The job of the team leader is to:

- remind the team of its schedule
- bring to their attention options and consequences
- solicit input from everyone regarding the next move
- help redirect the team toward their new goal

The team may have very good reasons for abandoning their agenda and discussing other topics. Sometimes, in a setting with new ideas, this kind of 'detour' is beneficial. It's okay to diverge from the agenda if the entire team agrees that is the best option at the time, and it is the facilitator's job to get that input.

2.12   Something often forgotten

With today's stringent diversity laws and regulations it is imperative that, as team leader, you keep notes and document all the team's decisions in case of disagreement. Otherwise it could be another member's word against yours. You might be sued! Record all the key points, ideas and decisions.

When taking notes, however, *don't edit*! Record comments verbatim as much as possible. At the end of each team meeting or one-to-one discussion, ask the team to email you the actions decided. This keeps your workload to a minimum and ensures you know that the team will do exactly what you request. *The team leader is responsible for documenting the team's process, discussions and decisions.*

2.13   Staying on track

- Set clear time limits for making decisions and remind people often of the time.
- Communicate decisions first and then give reasons why. It should never be the other way around.
- Explain the benefits of decisions you communicate.
- Seek team members commitment/buy-in to implementing the work to be done. Agree by consensus that everyone will accept responsibility for any extra work.

Remember this . . . Team members must commit to the success of the group and promise to participate.

Check your understanding of when to be a team leader or team facilitator by selecting true or false to the following questions.

Exercise

		True	False
1.	The team leader should make sure everyone participates.	☐	☐
2.	If it appears the meeting may go over time, team members should notify the team leader.	☐	☐
3.	Team members should accept the agenda as it is, and not make suggestions for modifying it.	☐	☐
4.	Team leaders should attend all meetings, but never say anything. It tends to intimidate people.	☐	☐
5.	It's a good idea for the role of team facilitator to rotate from member to member.	☐	☐
6.	Team leaders should paraphrase group decisions in order to get as much information as possible down on paper.	☐	☐
7.	Not-so-helpful team members are simply part of any team's make-up. Accept these individuals for who they are and let them have their say in team meetings. It helps them to get things off their chests.	☐	☐
8.	Team leaders are focused on process, not task.	☐	☐
9.	Team leaders are focused on task, not process.	☐	☐
10.	Team members should be empowered to discuss all details about team meetings with others in the organisation, even those who are not members of the team.	☐	☐

Feedback – you decide

Compare your responses with the feedback given below.

> 1. **The team leader should make sure everyone participates.**
>   The correct response is True. Ensuring participation is your responsibility.
> 2. **If it appears the meeting may go over time, team members should notify the team leader.**
>   True. Team members should feel free to speak up.
> 3. **Team members should accept the agenda as it is, and not make suggestions for modifying it.**
>   False. Team members should be encouraged to add input to the agenda.

4. **Team leaders should attend all meetings, but never say anything. It tends to intimidate people.**
    False. Leaders can and should attend early meetings and provide input. A leader's contribution should never be overbearing or dictatorial. After a period of time, leaders may not have to attend meetings.

5. **It's a good idea for the role of team facilitator to rotate from member to member.**
    True. Team members can gain valuable insight into the process of conducting an effective meeting when they rotate through various roles.

6. **Team leaders should paraphrase group decisions in order to get as much information as possible down on paper.**
    False. You should always record comments and suggestions verbatim where possible. The goal is not to write down everything that was said; that's not possible. Rather, you should summarise points, decisions and other items accurately. For example, if the team suggests, 'We need 10 new computers' you should not write 'More computers needed'. You should write '10 computers requested'.

7. **Not-so-helpful team members are simply part of any team's make-up. Accept these individuals for who they are and let them have their say in team meetings. It helps them to get things off their chests.**
    False. Disruptive team members, even those with good intentions and who are trying to help, should be dealt with appropriately, never ignored.

8. **Team leaders are focused on process, not task.**
    False. The team leader's primary responsibility is to keep the team focused on the task and accomplish it.

9. **Leaders are focused on task, not process.**
    True. Team leaders are primarily interested in completing the project, the final decision. Team leaders want a product, an item, etc.

10. **Team members should be empowered to discuss all details about team meetings with others in the organisation, even those who are not members of the team.**
    False. Team members should agree to keep important information confidential. What is important information? That's for the team to decide!

2.14   Chapter summary

- Being a team leader is rewarding and challenging.
- The team leader's role is to achieve the team's objectives.
- The team leader's specific responsibility role is to (proactively) control change.
- The great team leader ensures that his/her team does not have to deal with (react to) the same mistakes over and over again. If this happens, some manager, some place, is not doing their job.

Chapter 3
Leadership styles and required attributes

This chapter will explore the concept of leadership. It will argue that leadership needs to be considered in a broader context than our normal assumption of the word. It will then move on to show how team leaders need to balance their management and leadership roles. It will finally give you the opportunity to consider your key characteristics and attributes, and how you can use these to your advantage both now and in the future.

3.1   Introduction

Are you aware of the impact your management 'leadership' style has on others? Is it the most appropriate for enabling you and others to perform and achieve results? This chapter will provide you with a range of management and leadership styles and the ability to judge when it is best to use each.

The purpose of leadership is to get somewhere. This is as true of the 'leader' (or leading shoot of a shrub) as of a military commander or an executive in a commercial, voluntary or public organisation. 'Leadership' has three implications: objectives, a team of followers, and a 'contract' between the leader and those being led.

- Without objectives, leadership is a vacuous exercise no matter how enthusiastic and compliant the team. In one sense, the objectives come first, since even ad hoc and formally leaderless teams will explicitly or implicitly 'elect' a leader once they agree where they are and where they want to get to. See more about objectives in Chapter 5.
- Without a team of followers, the 'leader' is an individual contributor and no matter how highly motivated and successful, he or she can never really be considered a leader unless there is vision.
- Without a 'contract', activity is unfocused and progress is haphazard or not made and the objectives are not attained within the desired constraints (usually budget and time).

The 'contract' also sets limits on the style of leadership. A conscript army in time of war has very severe penalties for refusing to carry out instructions, which leads to a highly dictatorial style. Most commercial or governmental bodies have a range of disciplinary measures by which managers can impose their wills. However, labour legislation and labour mobility have forced civilian leadership into a more consensual mode whereby the 'contract' is made on the basis of mutual respect, shared understanding of the goals and agreement over the means to be used to achieve these.

3.2   Leadership

To be a successful leaders you have to radiate confidence and optimism. Dwight D. Eisenhower spoke about the great responsibilities he shouldered during World War II: 'Optimism spreads down from the Supreme Commander; pessimism also spreads down from the Supreme Commander, only faster'. Churchill took the same view and though he confided darker thoughts to his diaries, his message to the free world was, 'We will win. It will be hard ('Blood, Sweat and Tears') but the forces of good are on our side and we will win.'

These sentiments can perhaps be repackaged into one word: belief. The successful leader has to believe in the validity of the goals and plans, to believe in his or her own personal ability to lead the 'troops' to success and to believe in the ability of the 'troops' to overcome adversity and do whatever is necessary to achieve that success. Of these, self-belief is the greatest.

3.3   Leader or manager

Any organisation that wishes to shift its competitive positioning faces the difficult task of developing a leadership style and process which reflects the organisation's vision and helps managers and team leaders achieve that vision.

Our approach to leadership places the emphasis on leadership as *doing* rather than *role holding*; on expertise sharing rather than power-position deciding; on initiating rather than reacting to direction; on supporting others to perform rather than directing others; on team consensus rather than leader decisions.

It implies that the leader is not only the most senior executive but is also every individual in the organisation. *Leadership* becomes key, not who is *the leader*. Every time people take action to help make the organisation more effective they are demonstrating leadership. All members of the organisation need to be 'enabled/empowered' to do this.

In the past, the individual at the top of the organisational hierarchy was seen as the leader. To a lesser extent the members of the key executive team were also seen as leaders, but below this level other members of the organisation were not.

In a major contemporary transformation, all members of the organisation are now seen as potential leaders. It depends on the situation. Any member of the organisation no matter at what hierarchical level will have the opportunity to act as leader from time to time. Empowerment means enabling these people to adopt leadership when the opportunity presents itself.

Case study

A very good example of this is provided by Jan Carlzon, Chief Executive of Scandinavian Airline Systems, in his book *Moment of Truth*.[1] He describes a passenger arriving at the Stockholm airport check-in desk, to discover that he has left his wallet and air ticket in his hotel room. After explaining his dilemma to the check-in clerk, she took her personal power and leadership by:

- checking him in for the flight, on the basis that his name was on the reservation list
- phoning through to the hotel to have them check his room for the wallet and ticket
- arranging with the hotel to send the documents by taxi to the airport
- allowing the passenger to go through to the departure lounge anyway, because he explained that it was important he caught his flight.

In fact, before the flight boarded she personally carried the wallet and ticket through to the departure lounge to the astonishment and delight of the passenger.

Carlzon demonstrated two points through the story. First this is his idea of excellence in customer service. Second, and perhaps more importantly, the check-in clerk 'broke the rules' to give this quality of customer service. She had no authority to check someone in without a ticket, to hire a taxi to retrieve the documents or to allow a passenger through to the boarding area without having a valid ticket. To do all this she had to make a decision that on this occasion she would set aside the rules: in other words she took her personal power, her own leadership, to meet the needs of the specific situation.

This concept is about leadership more than about leaders. It is the taking of initiative to meet the needs as they arise. The notions that leadership resides at the top of the organisation is therefore no longer useful. Leadership resides wherever it is taken. People in organisations are increasingly being asked to take their leadership: to act with authority based on their knowledge and understanding of the situation they are in.

Modern organisations cannot run on the basis of the 'the great man' idea of leadership, that somebody at the top is the leader and all should look to this person to bring focus and direction to the organisation. The modern chief executive is as much a coordinator and vision builder as direction maker. Leadership must be shared by all members of the organisation. This means that managers in contemporary organisations have to think through their roles more clearly than ever before. There is a symbiosis between the roles of leader and follower which requires that the manager understand the times when it is appropriate to take leadership, and the times it is appropriate to take followships.

3.4   A team leader's dual role

The current trend is that most team leaders and managers need to be leaders and operational managers. This is due to flatter hierarchical structures and the need to respond ever more quickly to customer and market demands. What's the difference between leadership and management?

- management is **operational**
- leadership is **evolutionary**

Evolutionary leadership is one of the most important things a company, a division, or your department needs. In Chapter 2 we said that the manager's specific responsibility is to control change. Here we are saying that as organisations get flatter the team leader sometimes needs to think like a leader and at other times think like a manager. You must decide which leadership or management style is best. This, of course, depends on the situation you face and where in the hierarchy you are working.

3.5   The difference between senior leaders and team leaders

What must senior leaders (those at the top of organisations) do? They must establish the organisation's vision, strategy, culture and structure. This means that the most senior organisational leaders must generally forget concrete goals. That is for managers and team leaders only.

- **Vision** – Where are you going? What is the goal of the vision? This is not about concrete goals or objectives. For senior leaders the goal of the vision is competitive advantage. For us this is (or should be) usually about the image or perception employees, customer and stakeholders have of the organisation. The most important/difficult thing for senior leaders is to bring this vision or 'sense' into his or her team.
  - The leader's role is to communicate this concrete framework to them.
  - The manager's role is to think how to do it.

- **Strategy** – What framework (not tools) do you use to get there?
- **Culture** – What attitude should you have?
- **Organisation** – How best can you surround yourself with winners? How can you get your team to help one another? How can you know who is best for each job in the organisation?

3.6   The departmental team leader

Excellent organisational leaders can't do it alone. Neither can you, of course. You can try to do it on your own but a team of one is a lonely place and definitely no fun. While as team leader you are normally 'given' the organisation's vision to work with, you are certainly involved in creating and aligning your department's strategy with the organisation's strategy and vision. In terms of structure you are often limited by budget constraints, but your team's structure is filled by the people you chose. You, as team leader, must communicate the organisation's and your department's vision and strategies to all your team members and take day-to-day change to them. The knowledge each of your team members possess is an essential resource that must be understood and shared by everyone in your team. Crucial knowledge resides at the level of customer contact. Ensure that those at the top in your organisation's hierarchy hears this information, otherwise they may become out of touch.

3.7   The team leader as manager and a leader

Team leaders today are faced with a task of managing teams in a new and different ways. Many of the old problems still exist – how to motivate, what to control, how to measure, how to reward, what needs guidance, how to demand performance, how to structure processes. But in addition a whole new bunch of issues face the modern team leader – how to empower people, how to build a culture of high performance, how to balance minimal overheads with excellence in output and quality, how to create and maintain vision, how to operate effectively without many layers of management, how to maintain motivation which is not dependent on promotion, how to build careers where fewer layers of hierarchy restrict the promotional opportunities, how to manage without use of old style authority and with a much more educated and informed workforce.

All this means that today's team leaders need to be *both* leaders and managers. This is partly due to changes in organisational structure. While organisations are becoming flatter they are also moving towards lattice and cluster shaped. In lattice organisations, for example, lines of communication are direct person-to-person, often without intermediaries. There is no fixed or assigned authority; there are sponsors, not bosses, etc. In these organisational structures managers and team leaders need to exercise management and leadership styles of working.

What are the differences between team managers and team leaders?

The manager...	The leader...
administers	innovates
is a 'copy'	is an 'original'
maintains	develops
focuses on systems and structure	focuses on people
focuses on control	inspires trust
takes a short-range view	has a long-range perspective
asks how and when	asks what and why
imitates	originates
accepts the status quo	challenges the status quo
is a classic 'good soldier'	is his or her own person
does things right	does the right things

3.8   Traditional classification of management styles

What is the traditional classification of management styles and how do they impact on you as a team leader and on your ability to control and to communicate (Figure 3.1)?

* *Authoritarian* – An authoritarian management style means that the team leader keeps the initiative to him or herself by making all decisions, giving orders to staff and fully concentrating on monitoring the activities of others. It's all top-down one-way communication. The positive influence of the staff is minimal.
* *Laissez-faire* – A laissez-faire style means giving people a large right to self-determination. As a result they will develop their abilities. The team leader interferes minimally with the activities of his or her staff.
* *Democratic* – A democratic team leader aspires to achieve greater involvement of his/her staff in decision making. Collectively discussing a strategy to be followed and collectively making decisions (taking the firm's rules into account), a democratic manager monitors less. This person stimulates team members to self-monitoring and initiative. In democratic management there are two divisions: coaching and supporting. In this style there is clearly a strong overall 'helping hand' (not babysitting, mind you) by the team leader.

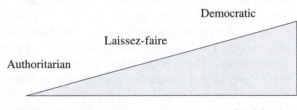

Figure 3.1 Classification of management styles

Which style is best? Often the democratic management style is considered to be the ideal. Is this correct? Let's look at the following situations:

- There's a fire at a shopping centre. The fire brigade sets out, and at the site the senior fire officer assembles his men in a meeting to decide together the most effective way of tackling the fire.
- In your department your manager wants to discuss and agree all stages of projects you and your colleagues are handling.

From these two examples it appears that the democratic style is not always the best. The fire officer should act in an authoritarian way and the department manager should give your colleagues ample room with the laissez-faire style.

A drawback of the classification into authoritarian, democratic and laissez-faire management is that these terms are only labels. A team leader who thinks of him or herself as a democratic manager may be considered otherwise by his team!

It is clear that the democratic style is not ideal in all situations. Good management depends on the development of a number of personal characteristics or attributes. Above all, management is not a static activity. Situations differ; people are different. The way 'management' is carried should also differ depending on circumstance. An effective team leader is flexible. When a manager/team leader adapts his/her behaviour to the circumstances, his/her 'style' is called **situational management**.

The basis of motivational excellence is good management. Good management is not a team leader who only influences the behaviour of team members once. You must *continuously* influence them. This is not an easy task considering that situations are never static. There is no one magic formula for the ideal style of management. Flexibility is necessary. Flexibility is founded on your ability to be *consciously* aware of your environment and your role in it. The point here is that if you are not aware of your style (or anything else for that matter) you cannot change it.

Consider the dynamics at play in which team leaders must operate. What influences are they exposed to?

- **From the team itself**
  Who are the people to be managed? What is their background, experience, age?
- **From superiors**
  What does senior management expect from you? What is their individual management style?
- **From internal/external customers and third party stakeholders**
  What demands exist?
- **Environment**
  What is the culture in your company like? What standards and rules are set?

Can situational management be better structured? The behaviour with which a 'situational-style team leader' approaches reports/members of the team falls into two categories:

- directing/task-oriented behaviour
- supporting/communication-oriented behaviour

Directing behaviour means the behaviour a team leader uses to inform or communicate with members of the team about what they must do, how they must do it and when they must do it. A good example of directing behaviour is the way a dentist might fill a tooth. When you had your first filling, did your dentist – without paying attention to your reactions – order you to open your mouth, open it still wider, tell you to clench your top and bottom teeth together and then to rinse your mouth out? This communication is all one-way. Even if you had been able to communicate, the dentist who uses directing behaviour will get on with the task, and periodically just check for levels of discomfort.

Supporting behaviour is characterised by two-way communication. This team leader stimulates, listens, shows understanding and puts himself or herself in the shoes of the person they are speaking with (rather than to).

Excellent management means being aware of the four styles. Which style to use depends on the situation. The extent of task – and communication – orientation is indicated in Figure 3.2.

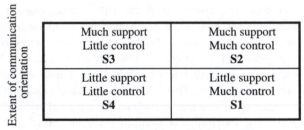

Extent of task orientation

Figure 3.2 Overview of the four situation-dependent management styles

3.8.1   Style 1 (S1): Authoritarian style

This management style is characterised by a strong appreciation of tasks. Work is there to be done; members of the team do the jobs they are paid for. This team leader gives instructions and supervises. Working conditions are such that there is little room for personal contact and interest in each other's problems. This management style is task oriented.

- people are paid to do their jobs
- strong supervision and control
- little room for personal contact with team members
- no interest in problems that are not directly related to the task
- one-way communication
- assigning tasks
- monitoring progress
- detachment

Used correctly
The surgeon gives instructions during an emergency operation. The patient is in mortal danger. In order to save the patient's life, the surgeon must make decisions rapidly. The surgeon instructs personnel in the operating room briefly and to the point and will scrupulously supervise the correct execution of orders given.

Used incorrectly
The group of technicians have thought through a variety of ways that they might use to present their project report. The team leader decides on an approach and instructs the group to use this approach with little or no input from those who had worked on the assignment.

3.8.2   Style 2 (S2): Directive style

The team leader gives a lot of instructions. At the same time he gives the members of the team plenty of room to ask questions, but is ultimately the decision maker. Furthermore the team leader clarifies why certain matters must be dealt with in a particular way.

- lots of instructions from the team leader
- room for the team member to ask questions
- two-way communication
- assigning tasks
- explaining reasons for and benefits of tasks
- involving the team members in working towards the solution

Used correctly
A new member of staff enters the office on their first day, bursting with enthusiasm. The team leader explains exactly what his/her activities comprise, how they must be carried out and why it is so important to act that way. At the end of the meeting the new staff member has the opportunity to ask questions on points not yet clear and check other points of interest.

Used incorrectly
The technicians are – without exception – very capable and experienced. At the end of their weekly team meeting, the team leader makes a presentation, including Q&A, on how to use new equipment recently purchased for the department.

3.8.3   Style 3 (S3): Supporting style

With the supporting style a team leader gives maximum attention to the team member as an individual. Motivation and a positive team atmosphere are key factors. The team leader has many formal and informal conversations. Tasks are given to the team. Sometimes orders must be given but always with reasons why and with explanations of benefits to the organisation, the customer and/or the team. With the use of friendly lobbying the team members are encouraged to deliver better performances. The need for control and supervision is minimal.

- maximum attention to the team member as an individual
- focus on positive atmosphere
- focus on motivation
- little importance given to control and supervision
- focus on dialogue
- two-way communication
- shows understanding
- supportive and helping

Used correctly
The senior technician felt overlooked when a new engineer was employed two weeks ago. The technician's results in the last two weeks have deteriorated considerably. The team leader meets with the technician and discovers his concerns about missing out on a promotion. In addition, the team leader asks the technician to join a senior working group which is to coordinate the improvement of the parts pre-ordering system.

Used incorrectly
A member of your team continuously scores poor efficiency results. He also frequently demonstrates poor product knowledge and makes incorrect repairs. He has been spoken to several times in an effort to improve his performance through extra guidance and agreeing on targets. Last week you were phoned by an angry customer. The technician had apparently failed (for the third time) to repair equipment. You decide to talk with the technician to learn what's going on and how you can help him to regain pleasure in his work.

3.8.4   Style 4 (S4): Delegating style

Delegating is the most important aspect of S4. The excellent team leader sketches in broad outlines what is expected from a member of the team, giving concrete reasons and benefits of decisions or tasks. How the task is to be carried out is left to the team members themselves. For experienced staff, supervision and involvement is hardly under discussion; only the final result is evaluated. The team leader assumes that staff will motivate themselves because they derive satisfaction from their activities and their contributions to the team. Communication is limited to what is absolutely necessary, the tone is businesslike and overall one-sided.

- team leader states the objectives but not how it is to be reached
- the team members must be creative and this motivates them in finding solutions
- one-way communication
- hands over authority while retaining responsibility

Used correctly

A team of scientists are at work in a research laboratory. Each is a specialist in their field. Their team leader indicates the conditions effective medicines must be developed in and what the available budget is. Research progress is discussed monthly.

Used incorrectly

A capable programmer is assigned the task of rewriting a program he is actually unfamiliar with. In the past he has proved insecure in new situations, which resulted in unnecessary mistakes. The team leader hands this task to the programmer on the last day before his own holidays and indicates that he expects to receive the results on his return.

3.8.5 When should you use which style?

It depends on two factors: (1) the competence level of your staff member and (2) the work situation.

The *level of competence* can be defined as the extent to which members of staff, given their ability and willingness, are competent at carrying out a particular task.

NB. You are not discussing someone's personality, their values, age, etc. You are talking about their ability to carry out a task. If the team leader wants to estimate the competence level of the team member well, he/she should divide tasks into sub-activities. These sub-activities should be compared with the 'abilities' and 'willingness' of the person concerned.

By 'abilities' is meant:	By 'willingness' is meant:
experience	motivation
knowledge	tenacity
technical skills	effort
communication skills	flexibility

Judge your team members' competence by assessing the following ability levels objectively:

- experience relevant to the job
- knowledge required for the job
- ability to solve problems on their own
- ability to monitor their own performance
- ability to set/meet deadlines and to manage time
- need to take on responsibilities concerning the job
- need to carry out the job efficiently
- persistence in carrying out the job
- pleasure in carrying out the job

3.9   Personal attributes. What kind of person are you?

No two people are alike. It follows that there are many different kinds of team leaders and managers. Indeed, wherever you look within an organisation – at whatever level of seniority – you'll see the same roles being performed by people in very different ways. Some of us excel at getting results but are not so adept at dealing with people; others are good all-rounders. Some of us talk easily and persuasively with anybody; others are reserved and prefer to make decisions from a distance.

Case study: sample job description

An attribute refers to a quality or feature belonging to or representative of a person. Most attributes and competencies are listed in your job description. Job selection attributes, based on typical leadership competencies in a typical team leader job description might include:

(i) Basic attributes – essential criteria

- minimum of one year's experience
- one or more good degrees, superior intelligence
- evidence of high performance against the organisation's leadership competencies and at least one excellent appraisal with their current employer
- highly motivated
- good spoken and written English
- mobile worldwide – requires mobility between countries as business needs arise

(ii) Basic attributes – desirable criteria

- ideally fluency in one or more languages
- some international experience (work/study/travel)
- worked in any function – experience of more than one is an advantage

(iii) Leadership competence – displays the following

Strategic vision
- takes an interest in the business's strategy and the wider industry context
- thinks beyond daily tasks, interested in emerging technology and consumer trends

Values communication
- sets a good example
- understands how the values apply to daily work
- if they have direct reports, seen as a credible team leader
- good listening skills

Commercial drive
- high energy
- committed to achieving results
- reasonably knowledgeable about specific business sector

Building organisational capability
- may have had some involvement in recruitment
- organises work team effectively
- uses formal processes and informal networks and gets things done

Customer commitment
- 'front-line' focused
- knowledgeable about customers (internal and external)
- works hard to satisfy both internal and external customers

International team leadership
- some understanding of the international dimension of the business
- will probably have worked, studied or travelled in different countries
- empathetic to other cultures
- aptitude for working well with others

The above may not apply to all engineering job descriptions. However, there is a good chance that at some point in your career you will be asked to demonstrate your personal attributes and competencies in these areas of working activity.

3.10   Applying attributes

So how can you personally demonstrate attributes and competencies? Your organisation's competencies represent another key developmental challenge for any team leader. They focus on those areas that your organisation believes are critical to achieving its strategic goals. Each attribute or competency normally consists of four components:

1. **Attitudes and beliefs** – What do I need to think and feel about this? For example, how can I deliver outstanding customer service if I don't feel passionate about satisfying customer needs?
2. **Professional knowledge** – Does my current knowledge limit my ability? What else do I need to know and how can I improve my expertise and understanding?
3. **Behaviours** – How do I act and behave to demonstrate my competence to others?
4. **Organisation measures** – Am I delivering results? What do I need to achieve?

Needless to say, as you progress within an organisation, commercial drive competency is of particular importance, so we would like to expand on this in more detail.

Example: commercial drive competency

1.   Attitudes and beliefs: you are likely to excel at this when you ...

- are passionate about results
- have an unrelenting desire to win
- treat company assets as if they were your own

- like to take calculated risks
- take personal responsibility and believe that individuals make a difference
- enjoy satisfying customers and relate to their needs
- enjoy working alongside business partners

2.   Knowledge: you will need to develop knowledge in . . .

Industry & economy

- Local market conditions
- Economy
    - customer profiles
    - marketing media
    - labour market
    - technology
    - potential partners
    - competitor activity
    - legislation

Business theory/best practice

- Core drivers of profitability
- Marketing best practice

The organisation

- Organisation strategy
- Best practice around the group in terms of:
    - market share
    - improving revenue per user
    - reducing cost

3.   Behavioural indicators: you demonstrate this competence when you . . .

Plan effectively

- construct well considered team plans
- anticipate likely effects of marketplace developments and adjust plans appropriately

Think and manage like an entrepreneur

- have the courage to make 'big moves' in the marketplace, stealing market share
- evaluate opportunities vs. risks
- manage P&L tightly

Forge strong partnerships

- identify opportunities for partnerships
- put together robust deals

4.   Organisation measures: your competence impacts the organisation when ...

People measures
- employees think and behave like true owners

Customer measures
- the organisation's brand is widely recognised and regarded
- you gain market share
- you have a great reputation for service excellence

Financial measures
- strong focus on the bottom line and results are ahead of plan
- tightly controlled costs

Community measures
- suppliers and partners put extra emphasis on their relationship with you
- company leaders are seen as opinion formers in local businesses

Typical key development activities

To realise the above points, your organisation may have some ideas about activities that might assist you in developing your competence. The example below comes from a major telecoms organisation and is based on their understanding as to how managers can help their team leaders to develop their competence in 'commercial drive'.

1.   Help your team members understand the real cost/profit equation of your business. Team members who know the actual value of the team's services/products and the margins that need to be maintained to drive profitability are less likely to ignore opportunities to reduce cost or increase revenue.
2.   Establish the results your team must attain as well as key milestones along the way. Use formal meetings as well as informal conversations to ensure everyone has their sights set on the same goals.
3.   Compare against 'best in class'. Your team's performance is only as good as metrics to which they are compared. Consider that you should look at 'best in class' information *across* markets, not just within the market.

Quick tips and ideas to promote commercial drive in your team

1.   Look for ways to reduce duplication of effort in your team.
2.   Engage in 'trend analysis' and explore possible improvements to problems identified.
3.   Initiate/sponsor a process improvement effort to eliminate duplication of work in your team or area using process improvement methods.

4. Identify the 'critical path' for achieving results. Plan the work, and then work the plan.
5. Have recommended reading available for each competency to allow your team members to research and study.

3.11   Chapter summary

- Management is **operational**; leadership is **evolutionary**
- Organisations need team leaders who demonstrate leadership qualities and can also manage
- In today's highly regulated but fast-paced business environment and with current diversity laws, organisations require throughout the hierarchy personnel who take personal responsibility for their actions
- Be aware of the classification of management styles into authoritarian, democratic and laissez-faire. These are labels. Good team leadership depends on the development and appropriate use of a *combination* of management styles
- Take into account your key attributes and how they might best be developed and used in a business context

Part 2

Creating the environment to make it work

Chapter 4

Continuous improvement

This chapter will explore the culture and environment that need to be in place for continuous improvement to flourish. It will look in very practical terms at how processes and technology should underpin your activities, and offer ten principles and a set of conditions for successful implementation.

4.1 Rationale for continuous improvement

So, having explored the roles, responsibilities, leadership style and attributes required to be an effective team leader, there is a final ingredient that needs to be explored: how to create the environment or culture for your team to truly flourish, and be motivated not only to be competent in what they do, but to be stimulated and energised to contribute more in terms of ideas for improvement. This could be with the service you offer or the products you produce. This will involve a review and awareness of attitudes, behaviour and work methodologies. It involves everyone in your team – including you – and needs everyone's commitment.

With the ever-demanding needs of the customer and current changes in the whole way that organisations are marketing their services and products, companies have to be far more agile and responsive to market trends. Where the market is competitive, the product lifecycle is short and speed of reaction is important, breakthrough project teams or task forces are formed specifically for the purpose. This involves bringing together the appropriate resources, the right people at the right level of an organisation, and letting their expertise flourish. Charles Handy describes this as a 'task culture'.[1] Individuals in this culture have a high degree of control over their work, judgement is by results, and there is mutual respect based on expertise and capacity rather than on status or age.

Organisations cannot thrive solely on this form of activity. There also needs to be evolutionary activity going on throughout the organisation based on the continuous examination of workflows, quality improvements and problem-solving activities. The Japanese, who excel at this form of activity, call this *Kaizen*. The word literally means 'change and good', which in business terms translates as 'continuous improvement'.

What do we mean by 'culture'? It is 'the way that things are done around here', or the way people behave towards each other in any business environment. In teams, the culture is usually determined by the team leader. We are not talking about the culture of your organisation here. We acknowledge that it will have an enormous impact on the ability of you and your team to be heard and to achieve your objectives. However, you as the team leader, can create a motivating environment where everyone is heard, nobody feels rejected or left out and everyone's opinion is valid.

Your responsibility is to align the informal team (culture driven) and formal team (management designed structure) to maximise efficiency and effectiveness. Team culture must be mobilised to support a formal team because, without the support of the informal team, the effectiveness of the formal team is diminished. Culture ultimately controls what, how, why, when and if things get done.

A person can make improvements if they *try*. They will only make improvements continually if they *care*. So it is your job to create an environment where team members feel needed, supported and valued as individuals. How do you go about this?

You cannot force people to come up with ideas for improvement. You need to create an environment where people have a sense of pride in their work, where they feel that their ideas and suggestions are listened to, that feedback is given, not only for ideas that are implemented but also when they are not.

Reward and recognition are obviously part of the equation, but also the right to take risks and experiment without fear of retribution if things go wrong. Personal development and multi-skilling are other vital ingredients that make people feel that your organization cares about them as individuals, and which in turn will encourage them to want to contribute more.

People in their private lives often show great creativity, be it in gardening, home decorating, or photography ... and yet how often does that creativity get transferred into the workplace? We do not believe that people come to work to do a bad job. It is your responsibility as team leader to create a team where people can grow and flourish and be creative. So let us look at what it takes.

4.2   The continuous improvement environment

4.2.1   Openness

Openness in a continuous improvement team culture manifests itself in two ways: through the infrastructure of your offices or work areas, and in interpersonal relationships.

Office space will usually be open plan, with individual offices only for the most senior members of management or for meeting rooms. There is a distinct lack of status symbols such as car park spaces being reserved for key personnel, or separate dining rooms.

Unisys developed this concept comprehensively and well in the form of Business Centres throughout the UK. In these Centres no-one has their own personal desk, but operates out of a large locker space. For routine work there is an open-plan area with

PCs and administration support personnel. The area immediately round the administration area is inclined to be noisy, as telephones are being answered and there is discussion around work projects. Should a member of staff require a quieter area for concentration, then they can book a PC at the furthest end of the room away from this area. For those who require total peace and quiet, for example if they are involved in writing a major tender document, there is a completely separate floor with individual offices.

Attached to these offices is another administration resource, which is able to source and research specific information required for personnel working on this floor.

In interpersonal relationships there is also a constant quest for openness. As the continuous improvement guru, Dr Deming, put it, success is about driving fear out of the team.[2] It is about causing everyone's opinion to be valued, and removing the threat from minority views. Within this environment there is an open acknowledgement that in everyday team activity problems exist. It is important that they are not swept under the carpet, or simply put through to the 'rumour network' to be addressed in a fire-fighting way. There is a need for root cause analysis to be undertaken to understand why problems have occurred in the first place, followed by cross-functional activity, if appropriate, to systematically address the issues. Persistence is the key, asking 'why' five times until you get to the root cause.

4.2.2   Harmony

As witnessed during the opening ceremony of the 2008 Beijing Olympics, the concept of harmony is of great importance in Eastern culture. When Sir Edmund Hillary and Tensing first climbed Everest in 1953, the papers in the USA and UK reported Everest as being 'conquered'. In Japan the news was reported as 'Everest befriended'. In a Western environment the achievement was seen as conquering against the odds. In Japan the mountain was seen as aloof and needing befriending.

This 'befriending, harmonious approach' transfers itself in Japan into team activity, and will manifest itself through non-adversarial communication and avoidance of interpersonal confrontation. This approach will often create frustration for Japanese team leaders when they are negotiating with foreign nationals or staff, since there are not only language difficulties to overcome but also cultural considerations.

Clearly, for those of us working in the West, a pragmatic approach should be adopted. The harmonious element, however, can be put into effect through sufficient cross-functional activity, secondment, or staff spending an agreed time frame in another 'internal/external customer' unit with which they regularly have to have contact.

4.2.3   Inform

The vision and strategies of the organisation need to be communicated on a regular basis, and must excite people by connecting to their values. The word 'inform' does not just mean to tell people information, but to put it into context. It is important to ensure that your team members not only understand the facts but also the implications and the context of the information that you are supplying.

Should you, for example, be presenting facts and figures, then your team may need basic financial training in order to be able to interpret the information. Statistics should also be produced in a user-friendly way.

Very often we visit organisations that display their results for staff to see. The statistical information will often be produced in figure format, and laid out on paper that has started to curl over and look like tired cheese sandwiches! On a client site that we visited in the USA, the president produced all the pertinent information with regard to shipping, distribution and production in graph format and put small personal observations against some of the pie charts. This brought all the information to life and made far more sense to the staff reading it.

Information is rather like a lift. It only works well if it goes up as well as down. It is the team leader's responsibility to ensure that the information doesn't come up through filters and down through megaphones.

4.2.4   Learning environment

In a continuous improvement culture everyone – not only senior or middle managers – is involved in the learning process. Training will be as and when required rather than on a blanket basis, and will often be undertaken at a more localised (monthly team meeting) level by a team leader/facilitator. Education will not only be task related but also include communication skills and team-building activities.

Training will often be the first thing to be cut during periods of recession in business. Organisations such as Nissan who adopted the principles of continuous improvement have continued to train their staff even when there is a downturn in the economy.

A learning environment encourages every form of personal development, and organisations such as Mars Confectionery and Shell have learning resources in place for many forms of leisure activities.

'In the simplest sense, a learning organisation is a group of people who are continually enhancing their capacity to create their future. The traditional meaning of the word *learning* is much deeper than just *taking information in*. It's about changing individuals so they produce results they care about, accomplish things that are important to them.'[3]

Culture change to introduce continuous improvement is not a quick-fix programme but a slow evolution towards desired behaviour. However, should certain personnel consistently inhibit the introduction of the 'new' culture then they need to be supported by the appropriate methodology, be it through mentoring, training or some form of supportive development programme. If there should then be little or no change in their behaviour, they will need to be removed from the team. Scepticism can be rife during any improvement programme, and unless people can feel and see change occurring, they will wait until the new initiative 'blows over', and in the meantime see how colleagues are siding up and take a 'political allegiance'. Jack Welch (GE) believed that behavioural consistency with organisational values matters as much as task accomplishment. He cited organisations that have made the tough call and exited individuals who do not live the values, even when their financial and/or operating performance is well above average.[4]

4.2.5   Enablement/empowerment

True enablement/empowerment cannot happen unless people have been given the tools to do the task or job effectively. In many instances comprehensive communication of the objectives is sufficient, but often people will need training, be it on or off the job, in order to move forward. Delegation with authority is vital, plus encouragement, feedback and reward. This is such an important factor that Chapter 10 is dedicated to the topic.

4.2.6   No blame

If something should go wrong within your team, it is important not to automatically assume that the person who is undertaking the task is at fault. It might be that there are inhibiting factors within the organisation or team itself, poor cross-functional links, for example. It could be that your initial instructions were not as clear as they should have been, and you have not offered sufficient support during the project for that person to be able to undertake the task or responsibility appropriately.

Within your department good news should be accredited to an individual in the team. The group should share bad news, and everyone should trust you as a team leader to look after them in a disaster. It is important to give credit where it is deserved (you get reflected credit), and shield your staff from disasters. People clearly have to learn from their mistakes but you need to take issue with that person behind closed doors. Part of this process is to constantly create expectations for your people to aspire to. In continuous improvement there is no such thing as the status quo.

A recent report by the UK Government's Department of Trade and Industry found that many organisations claimed to have a no-blame environment until something went wrong and then everyone looked for somebody to blame! The way to ensure mediocrity is to put people at risk, then blame them for their failings; guess what, they won't expand and take risks for fear of failing. Bill Gates claims that he will not employ a senior manager unless he or she has made mistakes.

4.2.7   Making constant improvements

Searching for new and different ways to undertake a task or activity is an essential element in the creation of continuous improvement. The assumption is that everyone has two jobs: doing the job and improving the job. A new performance standard will be created through the improvement; however, this performance standard will only be in place for as long as it takes for another individual or team to find a better way of doing it.

Performance monitoring would appear to be the obvious approach to adopt, but as Masaaki Imai has put it 'Yes, it is common sense, but not common practice!'[5] In a traditional business, re-examination is a project. Within an organisation or a team that has adopted continuous improvement it is a whole way of life.

Many people associate continuous improvement with being applicable to a production environment and linked to monitoring systems, JIT, reducing waste, etc. This is a BIG mistake. Continuous improvement is all about culture. This culture must underpin everything you and your team do. Informing, no blame, and making constant improvements should become part of your team's daily approach to work.

Case study: L'Oreal[6]

L'Oreal, the international cosmetics giant, receives accolades for its health and safety standards. Achieving more than $4 billion in 2006 sales and with 7,500 employees, its six US plants receive recognition for implementing safety and health procedures through the Occupational Safety and Health Administration's Voluntary Protection Programme.

Every year L'Oreal challenges all of its factories with some internal goals on improvement. The factory comes up with its own safety, health and environmental action plan each year for improving its performance.

The culture of the organisation is based on key concepts of continuous improvement – where genuine concern for co-workers at all levels is fundamental to the way staff work together. All employees feel responsible for their own safety and the safety of others. The concept of giving back and sharing knowledge with others is also encouraged. Staff at several of the sites in the US act as mentors to business communities locally or at a regional level. L'Oreal is now a flagship for several work sites for agencies such as the US Postal Office and manufacturing plants like a Kentucky site for the General Electric Co. to elevate safety initiatives there and among smaller businesses.

4.2.8   Customer focus

Underpinning continuous improvement is a long-term focus on your team's customer needs. A quality product and quality of service are inextricably bound into the manufacturing–selling chain. Teams generally claim that they adhere to the concept of 'quality'; it is your route to promotion and team success. What is now required is a need to be memorable through the added-value service that you and your team offers.

So how can you constantly instil the importance of the internal/external customer in your team? Every visual aid should be used to highlight your customer focus, be it through internal team notice boards, on your stationery, or by circulating customer success stories via email.

According to the *Harvard Business Review*, if you can reduce customer defections by just 5 per cent you can increase profitability by between 25 and 85 per cent!

The exploration of every way to add value to the services that you offer should be a journey without an end, and the whole ethos of your team should be geared to encouraging this.

Case study: Cougar Automation[7]

Winner of the Manufacturing & Engineering category
Sunday Times Customer Experience Awards 2007

This manufacturing and engineering company designs and supplies automatic control systems.

Through differentiating the service they offer to clients they have managed to grow their net sales in the last financial year to £2.4 m compared with £2.3 m the previous year. Net profits jumped from £1,097 two years ago to £42,548 in the last financial year.

So how have they managed to achieve this?

They have moved away from the typical pitch in their industry – 'our engineers are better than yours' – to give the message, 'we are as good as anybody else, but the real difference is the engagement and enthusiasm we bring'.

Customer service starts before the contract is signed, by ensuring that their customer has the same picture of the outcomes as they do. Customers are involved from the earliest stage of system design. They have a flexible ordering system. Systems are tested by a different engineer, rather than the one who configured the system. And as staff are multi-skilled there is continuity throughout the project.

4.3 The bigger picture

Looking at the bigger picture, what else needs to be in place for continuous improvement to succeed?

It is vital that the team structure forges the shortest possible chain of command. The structure needs to be an inverse triangle with all activities being focused towards the internal/external customer.

For all forms of team activity there needs to be a person at a more senior level than the team leader to act as the team's sponsor. We like the American analogy; it is the person who 'rides shotgun for the team'. In the days of the Wild West, stagecoaches carried a person whose responsibility was to scan the horizon for menacing situations, hostile Indians, robbers or whatever. This person – presumably armed with a shotgun – was also expected to deal with these incidents. He (while we are no experts, we assume there were few, if any women so engaged), had no responsibility other than to ensure that the stagecoach operated smoothly, did not run into trouble, was defended against attack and so on. It has been my experience that this 'shotgun role' is critical not only at the setting-up stage of the team, but also in a supportive role during its activities.

4.4   Implementing continuous improvement

Having looked at your team's culture you now need to consider your responsibilities towards the resources and means to take action your team requires.

There are three key elements that need to be considered. First, you need the commitment of your team members as outlined below. Second, technology needs to be in line with requirements. Third, your systems and procedures should be regularly reviewed (Figure 4.1).

No part of this tripartite arrangement can work in isolation. You can have the greatest systems, procedures and technologies in place, but unless you have your team members 'on board', continuous improvement cannot develop. Likewise, you can have your team behind you all the way, but if they do not have the appropriate technology or systems/processes and procedures in place to support them, they will not be fully effective.

Looking at processes and procedures, one of the key concepts of continuous improvement is the elimination of waste. In Eastern organisations where they have adopted continuous improvement, they have what is known as the 'morning market' at the beginning of each shift or working day.

The purpose of this meeting is to examine the 'defects' or issues from the previous production run or day, and look for ways to improve the manufacturing or working process. In an administration function this will take the form of a weekly meeting

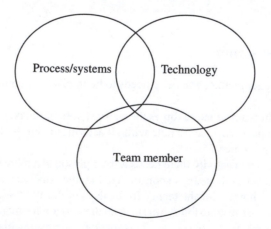

Figure 4.1   The three key elements

to review all the suggestions for improvement that team members have made, which of these ideas can be put immediately into effect, and which ideas should be taken forward to a cross-functional meeting stage.

Measurement of waste is just as applicable in the service sector. During a training session that we ran in Singapore we met a group from Ascott International who had realised a 7 per cent increase in income, a 5.3 per cent increase in employee productivity and a 6 per cent increase in profitability. These savings were achieved over two years through the elimination of waste, which they told the group could only have happened as a result of a total cultural change to adopt the principles of continuous improvement.

Here is a classic tool used at the start of any waste management review. It is called the 'Five S'.

The 'Five S'

- **Sift** (Seiri)
  Group items into those you need and those you don't
  Surplus in storage

- **Sort** (Seiton)
  Configure all essential items in work area
  Frequently used items close to hand
  Storage items – visually managed

- **Sweep** (Seiso)
  Clean and tidy working environment and equipment – enabling problem areas to be identified immediately

- **Standardise** (Seiketsu)
  Create a standardised system – well defined and documented – followed by all

- **Sustain** (Shitsuke)
  Five S as a way of life – established rules
  Management lead by example

This tool is widely used in the manufacturing sector, but there is no reason why it cannot be used in the service sector or office environment. Think how you slow down during the working day – searching through endless files and icons on your PC to find vital information that you need to use immediately. It is very easy to get into information overload with so much data to sort through, so you cannot see the wood for the trees!

The answer to this is to set up a regular review of all your electronic data. This could be on a monthly basis where, for example, any out-of-date reports or proposals that are not templates and have not been accepted are put into a 'dormant' or 'to be reviewed' folder. Set up automatic rules: if the data has not been accessed in, say, a three- to six-month time frame (depending on how you envisage the usage) then the it will automatically turn blue. During your monthly review of data you could then decide what you should do with the blue items, e.g. archive or delete them.

You will probable find that it takes discipline to get into the mode of doing this on a monthly basis, but once you start to see the benefit of reducing the amount of electronic data you are storing and the increased ease of access to current pertinent information then you will be delighted. Words and information clutter our lives. Keep it simple!

There are ten key principles you need to bear in mind during your development activities:

Principle 1

Continuous improvement is a series of balancing acts. Yes, be obsessive about developing new ways to reduce waste, but make sure your solutions fit into your long-term vision. The way to build new products/ideas is in self-governing teams who have demanding objectives. This is about making your team proud of itself. You must get your team members to listen to, but not always follow, what your customers say.

Principle 2

Start by deciding what you want to achieve. You must link this back to whatever 'winning' means for your company.

Principle 3

The essential resource in teams today is not capital but knowledge. This knowledge resides lower down in your team. Many teams recruit first-class people, do a great job in training, only to constrain team members with a highly authoritarian style of management! This won't work in a continuous improvement culture.

Principle 4

How often do you really communicate with all your team members? Do you take continuous improvement to front-line people? Do they feel that they 'belong' and are making a positive contribution to the business strategy of your team and organisation? If the answer is no, change this immediately.

Principle 5

Only customers can guarantee employment. All that companies and your team can offer is employability.

Principle 6

Your team members are the critical and ultimate source of competitive advantage for your organisation. Successful continuous improvement companies are obsessive about what they produce or the service they provide. Ensure your products or services become the glue that keeps your team together. Do your products or services dominate team corridor gossip?

Principle 7

Today's paradox: just as employees can expect less loyalty from companies, organisations are more than ever dependent on high-level performance from employees. This has an impact on continuous improvement.

Principle 8

In terms of continuous improvement, your team's vision is the idea of how your team's 'market' will look (or could be made to look) in future, which inspires and focuses team members.

Principle 9

The one essential question for team leaders is: *how do you architect a strategy that is aligned with organisational business priorities and capable of rapidly adapting to a shifting competitive landscape that will help us survive?*

Principle 10

Today's key core competence is not core *technical* competence but rather core *cultural* competence which encourages flexibility, change, learning and adaptability to customers.

And finally, what are the 10 conditions for successful continuous improvement implementation?

1. Demonstrate the need for the improvement.
2. Make the improvement integral for your organisation/team.
3. Use a step-by-step approach.
4. Ensure a fast execution of every individual step.
5. Ensure irreversibility.
6. Ensure sound management of the project.
7. Reserve, in advance, the personnel and financial capacity that are required.
8. Ensure broad participation of your team members.
9. Make sure there is a balance between substantive and process improvements.
10. Take enough time to implement the improvement.

Case study: Virginia Mason Health System[8]

Virginia Health System's vision to be a quality leader in healthcare meant continually thinking in new ways.

In the past they had relied on traditional methods to determine where defects occurred, using quality assurance processes, hiring the best nurses and physicians who came from the best training programmes; they formed a best practice task force that identified patient safety and quality issues. Senior management realised, however, that these measures were not enough to overcome an unreliable system that could not always identify or track where defects occurred. Having explored a variety of options in the health care arena, they opted for a system from the manufacturing sector – the Toyota Production System (TPS). Senior management visited Hitachi and Toyota factories in Japan, and undertook extensive training in the use of TPS tools.

The objective was set to find ways of streamlining repetitive and low-touch aspects of care delivery, in order to release staff to spend more time talking with, listening to and treating patients. Each hospital then chose a handful of departments to test-case the so-named Virginia Mason Production System (VMPS). Waste, or anything that did not add value to the patient, was eliminated from the process. For example, in surgery ten different trays were used for ten different physicians performing laparoscopic surgery. The new streamlined process standardised the trays, preparing only one. This was saving money, eliminating redundant processes and reducing human error.

Also, just as production can be stopped on the factory floor at Toyota when something goes wrong, staff were able to signal a patient safety event had occurred by email or phone, be it as severe as administering too high a dose of medication, or as trivial as a spill on the floor that might cause someone to slip. Management now respond immediately, or the process is stopped if it cannot reliably be assessed or fixed.

For a medical environment this change is a cultural shift in attitude. Nurses, for example, may now report concerns over the operating rooms and have the situation reviewed and responded to. In the old hierarchical structure concerns might not have been reported to management directly.

After six years of using VMPS Virginia Mason has experienced a big payback through improved patient satisfaction, better access to physicians, greater safety and quality, increased satisfaction and more efficiency and productivity throughout the organisation.

Some of the specific results include:

- a saving of $11 million in planned capital investments
- reduced inventory costs by more than $1 million
- reduced staff walking by more than 60 miles per day
- reduced labour expenses in overtime and temporary labour by $500,000 in one year.

4.5   Chapter summary

- Continuous improvement needs to be part of your team's everyday activities.
- Unless you get the culture right it can't happen!

- Openness and supplying broader context information should underpin your environment, plus learning and development should be there for all, and actively promoted.
- Enable, empower and support team members as they need.
- Do what it takes to create a harmonious setting through getting your people to be internally and externally customer focused.

Part 3
Practical implementation

Chapter 5

Operational management

To start this section on practical implementation we will first look at how a team leader needs to assess the use of his/her time. We will then move on to practical operational issues in terms of project management and how to run effective meetings.

Business practices are placing ever-heavier demands on team leaders. It is important not to allow these demands to overwhelm you. You need balance in your working life between the things you want to do and the things other people want you to do.

Team leaders must manage. You must learn better ways of using your time and also ensure that these practices filter down to members of your team so they become more effective in working towards departmental and corporate objectives.

So how can you plan the work, and work the plan? What is it that stops you being as effective as you would like to be? Procrastination – does it rule your life?

5.1 First things first

In your everyday working life there never seems enough time to undertake everything you wish and need to do. The reality is that you cannot invent more time; you need to find out how to use it more effectively. Pressure always seems to be more towards speed than excellence, something that prompted the author Robert Heimleur to observe, 'They don't want it right, they want it Wednesday'.[1]

Habits are powerful. Ones that need changing may take some effort to shift, but once new ones are established, then they make the approaches they prompt at least semiautomatic. Getting to grips with managing your time effectively may well take a conscious effort, but by establishing good working habits the process gets easier.

The reality is that people in general are good at procrastinating. This could be, for example, because you are unsure how to tackle a particular assignment, dislike the task or prefer another (despite the clear priority). Kate Muir in the *Sunday Times* supplement wrote a heartfelt article about procrastination.[2] The premise was that she personally loved deadlines as she is by nature an 'arousal procrastinator' – which, according to the online magazine *Slate* (www.slate.com) is a person who 'seeks the excitement and pumping stress hormones of having to finish everything under duress'!

Well, if you don't fit into this category, as many of us don't, a little conscious effort and planning can make all the difference to your life.

So as a starting point, here is a five-step process to help you become more organised and focused in your everyday activities.

Step 1: List all tasks and activities

The tasks or activities will vary enormously depending on your job. Some will be quite major and may require a fair amount of time. Others will be small and will not take long to do.

Step 2: Decide priorities

- There will be proactive tasks which contribute towards the achievement of the job purpose. These may not be urgent, but they are important.
- There will be urgent tasks, when you have to react to situations. Some of these will also be important because they contribute towards your longer-term goals. However, many of these reactive tasks, while being urgent, may also be relatively unimportant. They must be done, but not at the expense of the important tasks.
- There will also be other tasks which are both relatively unimportant and of lower urgency.
- There are some other things that should never be done at all.

So, is the task urgent or important?

- The urgency of a task should suggest *when* it should be done.
- The importance of a task should suggest *how long* you should spend on it.

Importance		
Low	**High**	
Category B1 High Urgency Low Importance	*Category A* High Importance High Urgency	**High**
Category C Low Importance Low Urgency	*Category B2* High Importance Low Urgency	**Low**

(Urgency)

First, consider every item on the list of tasks or activities. Decide which category each task or activity falls into. Having done this you can now clearly see:

- when things have to be done
- how much time should be allocated to each task

Step 3: Delegate when appropriate

- Having decided what needs to be done and the order or priority, the next question is, 'Who should do it?'
- In some cases the answer will be 'Me' simply because there may not be anyone else who can do it. However many you have in your team, there still can be situations when you are the only person available.
- However, in many situations it may be possible to delegate work to members of the team.

Delegation matrix

Importance		
Low	**High**	
Category B1 Delegate, get it done soon	*Category A* Do it yourself, now	**High**
Category C Delegate, keep an eye on it	*Category B2* Delegate, schedule time for completion	**Low**

Urgency

Category A: high importance–high urgency

- Do it yourself now – spend as long as it takes.

Category B1: high urgency–low importance

- Consider delegating as a development opportunity.
- Get it done soon.

Category B2: high importance–low urgency

- Consider delegating parts to an experienced and proven member of the team.
- Schedule time for completion.

Category C: low importance–low urgency

- Delegate, have it done sometime.
- Keep an eye on it.

There are going to be certain activities such as meeting with colleagues, opening mail, responding to emails, that are not so time critical but still need to be taken into account. In planning your diary you need to put in the time-critical activities and deadlines, but also schedule in the non-time-critical tasks in the gaps. So in brief:

Step 4: Plan your time

- Prepare a daily 'To Do' list, ideally at the end of the previous day.
- Schedule time for Category A high-priority tasks.
- Allow time for reactive tasks.
- Plan sufficient time for regular tasks that you know about mentioned above.
- Schedule time each day for planning.
- Block out time and batch together reports to read, or performance reviews to approve, to maximise the use of the mental set-up time needed for the task.
- When is your prime time? It is not always first thing in the morning; it could be a little later in the day. We all have different 'creativity rhythms' – find out when yours is and plan accordingly.
- Write things down instead of relying on your memory.
- Set deadlines and give yourself 'rewards' when you meet them.
- Have a visible, high-impact reminder of your job purpose to keep you focused on what's important.
- Improve reading skills – scan faster.

Step 5: Stop time wasting!

- Can you really afford to read junk mail?
- Handle each piece of paper only once.
- Start and finish one job at a time – avoid 'butterflying' between different assignments.
- If someone pops into your office and you don't want to be disturbed, stand up and walk towards them – they won't think to stay.
- Optionally, put your desk in a position that means your back is to the door – much less welcoming!
- In turn, of course, be conscious of wasting other people's time.
- Be more assertive – learn to say 'No'.
- Keep asking yourself, 'What's the best use of my time right now?'

'To manage time well you have to *believe in your own knowledge.* If you know a weekly meeting takes thirty minutes, don't convince yourself that today it will only take fifteen minutes simply because you have more to do. If you have to be somewhere in ten minutes and you have ten minutes to get there, don't make one more phone call simply because you want to get it out of the way. People who manage their time badly seem to want to be unrealistic and go out of their way to create out-of-control situations.'

Mark McCormack[3]

Reviewing your working activities will almost certainly highlight the two key activities in your day:

- project-related activities
- attending and running meetings

So now we will move on to looking at these topics in more depth.

5.2   Project-related activities

Our purpose here is not to explain the salient points or minutiae of managing projects. We realise that most of an engineer's, indeed a team leader's work (as Tom Peters said), is 'project work'.[4] Therefore, we will concentrate on the essentials of delivering successful projects in collaboration with team members, looking primarily at the behaviour and the motivational activities that lead to success.

Countless projects significantly overrun their schedule and budget, and as a consequence fail to achieve their organisation's financial and strategic objectives, often with sizable increases in costs and substantial financial losses. *Why?* This is due mainly to the failure of many project managers or team leaders to successfully apply the tools and techniques of modern project planning, monitoring, control and progress reporting on their projects. In addition to the financial losses suffered by the organisation, many such projects also fail to deliver the required quality of outcomes as a direct consequence of inadequate monitoring and control.

5.2.1   What is a project?

As Bonnie Biafore says in her book *On Time! On Track! On Target!*, 'a project is a one-time job with specific goals, a clear-cut starting and ending date, and – in most cases – a budget'.[5]

An internal or external client may well need your project, product or service, but also needs to satisfy personal objectives such as security and wellbeing. By understanding what motivates people and consciously working in harmony with them, you will become more effective as a project team leader.

5.2.2   The objectives of the project process

Every business project has objectives, otherwise there is no sense of direction and no way to measure success. This is as true of the project process as of any other. So what are the objectives of the project process? The primary objective is obviously to get the job done. The project process includes:

- set up: initiation
- set up: definition
- delivery planning

- delivery
- closedown and review

For many project leaders this is sufficient.

Now consider the project process from the viewpoint of your employer and of your direct manager. They are concerned with two measures: performance results (or value) and cost of the project. The latter can cover a wide range of expenditures, but a narrow version excluding general overheads is simply the cost of the project team members. If the project team members work effectively and easily then the cost per unit of project volume or value goes down, and management are happy. Let's call the ratio of project amount to cost of the project the 'efficiency' of the project team. It is obvious that if the project process can be made easier then project efficiency should rise and, conversely, if the project is made harder, efficiency will fall. We remember reading that typically 52 per cent of all projects finish at 189 per cent of their initial budget. Some, after huge investments of time and money, are simply never completed.

The objectives of the project process are therefore to maximise results while maintaining a high cost (time and resources) efficiency.

5.2.3   Raising project efficiency

While there are many factors that make running a project easier or harder, most of these are under the project leader's control. Project design, resourcing, sponsorship, project implementation and delivery are all controllable, while other aspects such as budget and delivery time may not be. So what can the project team do to make a difference?

In our experience, project team members have a high project efficiency for two reasons:

- They make it a pleasure for others to engage with the project team.
- They make it easy for others to engage with the project team.

The project team enlarges its 'circle of influence' because they treat themselves and others well. Project efficiency is raised if you can make colleagues and clients into active references. Assuming that your project suits the client's needs:

- They will 'self-refer' back to you when they need more.
- They will recommend the team to others.
- They will often be willing to recommend the team to third parties.

All three forms of active reference make your life easier and raise your project efficiency by extending your circle of influence.

It follows that the objective of a successful project team is to go one step beyond realising a result by creating active references. To do this the project leader must build and maintain a strong personal relationship with others. This is in part a purely personal thing, initiated when you first contact stakeholders, and in part a consequence of the level of professional service you provide both before and after the project.

So let's look at this relationship-building process in more detail.

5.2.4   The end-users view

Here we consider a few aspects which are common to all projects: why were you selected to lead the project? If you have an understanding of the end-user's motivation you are able to guide the project process in favourable directions and away from potential pitfalls.

The desire for security

Fear is a familiar primal instinct. It is of great value since it warns us of danger and sets in train a group of physiological changes which enable us to respond swiftly and effectively, the 'fight or flight' reaction. However, fear can also be counterproductive in that it can cause paralysis so that we do nothing. Fear of the unknown, fear of the future or fear of change come into this category since, in our business lives, change is continuous, and we all know that we have to accept change if we and our organisations are to prosper. We counter this sort of fear by seeking security.

Our ancestors did this by constructing familiar, safe places within stockades, castles and covered wagons; we do it with organisations and regulations. In a project situation we foster a feeling of security by providing reassurances. One set of reassurances relates to the end-user:

- The project outcomes will work and will satisfy the users needs and will be reliable.
- Your project team has a good reputation.
- The project team's service levels are high.

The project team's primary objective is to realise the project's objectives; the team leader's primary objective is to realise the project's objectives and develop positive relationships so that your end-users become active references. This is done not so much by taking the project team's or the project end-user's side, but by really listening to and eventually fully understanding the end-user's problem.

You do this by skilful questioning. You draw out everything you need to know and regularly recap your interpretation until the end-user or the project sponsor is secure in the knowledge that you and your team are in complete agreement with them. Your family doctor is a good example of someone who is able to make you feel secure by using a good questioning technique to discover all your symptoms and then prescribe a suitable treatment. A good project team leader is a sort of company doctor, 'curing' the project's ailments.

The desire to be noticed

When we go to work we exchange our time and skill for a variety of rewards. We may take for granted the pay cheque and associated financial or quasi-financial benefits such as health insurance. Many of us also go for the social benefits of spending much of the day in the company of pleasant, intelligent, interesting people. There are also intangible rewards which boost your project team's morale. The esteem of your project team members and sponsors is one of the most important. Everyone likes to be noticed and to have their achievements recognised.

The desire for wellbeing

This is another primal instinct. We shy away from situations which are physically, intellectually or emotionally uncomfortable, and seek out those which are comfortable and which give us pleasure. We choose our friends from amongst a different, even if overlapping, group of the colleagues who are chosen for us. We look for homes which are convenient for the shops, schools or for the journey to work. We fill our homes with goods designed to make our lives easier and more comfortable.

Most projects satisfy equivalent needs. While the end-users as individuals may not gain much additional wellbeing from the project, the project will have been designed to improve the efficiency, productivity or quality of life of others, and the end-user can take a vicarious sense of satisfaction from the enhanced wellbeing of his or her department or the organisation as a whole.

5.2.5   The major elements of projects

Gathering information

Great projects are all about understanding end-users' needs and motivations. It's great if you've been given a terms of reference but you will still need to ask questions so that you can understand who will benefit from the project, how they will benefit, maybe how it will interface with other projects and any other relevant information.

To give a simple example: when you walk into a garden centre and say 'I want to buy a spade', that is certainly clear, but the information is insufficient. The salesperson should want to know:

- who is going to use it (a professional gardener can handle a bigger, heavier implement than an unfit amateur; tall users need longer shafts)
- what for (a big blade for digging over large clear areas, a small one for work amongst the shrubbery)
- what sort of soil (stony or heavy clay soils suggest a strong narrow blade, lighter soil a wide one)

Then you can talk about the budget you have, and only then can the salesperson point you to the two or three most suitable spades in stock, and you can base your final choice on personal preferences such as materials or shape of handle.

This example illustrates that project team leaders should know a lot about their subject and understand how the various products/services match different user requirements. If you are to make rational recommendations you will need to find out all the relevant information.

Fortunately, people enjoy being the centre of attention and most people enjoy talking about themselves or their problems, so ask 'open-ended' questions to encourage them to open up. Intersperse purely factual information with opinions and subjective material which will allow you to understand something of their motivations.

Listening

Listen carefully. This is actually quite a difficult art, since you have to hear what they say, convert it to a mental picture, make notes, if possible draw diagrams which the prospect can agree or correct, all the while listening to the next sentence. The process is comparable to simultaneous translation, where specially trained linguists are able to take in a stream of words, translate these and transmit without losing the original speaker's flow.

Three things to guard against are:

- thinking faster than the speaker and missing important details because you are busy figuring out how to respond to the last thing they said
- filtering out things you don't understand or don't want to hear and so getting only half of the message
- simply mentally wandering off at a tangent

You know from your own experience that it is immediately obvious when someone you are speaking to has a lapse of concentration. The message you receive is that they are bored or that they have simply switched off, and so you switch off too. Rather than risk this, if you are having problems, interrupt (politely of course), explain that they are going a trifle too fast and ask them to check your understanding of where they have got to. Then you restart from a firm foundation and you have gained a small 'plus point' for honesty.

Don't be afraid to ask for more detail or additional explanation if you don't understand something. After all, you need to have a full understanding of the end-user's or stakeholder's problems that the project is intended to resolve. Keep an especially careful watch for any expressed dislikes. There must be a reason why your end-user/ stakeholder says 'I don't like X' and it is vitally important for you to know this.

Defining your project terms of reference

Your terms of reference is akin to your sales proposal. Project terms of reference, besides containing an overview of the project, clarify the issues to be addressed by the project.

Construct your terms of reference by including:

- information about the business problem (the needs the project is to address)
- information about the end-user's or stakeholder's 'desires' and 'wants'

- concrete knowledge of their current situation and the future 'desired' situation
- any special terms and conditions that must be considered during the project cycle
- the purpose of the project, and timed deliverables

With major projects the terms of reference is often distributed to quite a large audience. If at all possible, check with someone beforehand to ensure that the document contains the right 'tone' to appeal to the final audience, and to pick up on any omissions or unclear points. A mainly technical audience will expect quite a lot of technical depth. A mainly non-technical audience, on the other hand, wants just an overview of the technical aspects and will be more interested (and persuaded) by a business-oriented presentation which concentrates on showing them how the project will help them to overcome their business problems and become more effective in meeting their objectives.

Pascal, a famous French philosopher, mathematician and physicist (1623–1662), said, 'We are more convinced by the reasons which we have found ourselves than by those given by others.' For project sponsors to sign off on the project they must say 'yes' to each step of the terms of reference, because they recognise their own words in the document, i.e. their own evidence and reasons needed for the project.

At the end of the project

Say 'Congratulations. Thank you', ending on a positive note.

You have all achieved your very different objectives; you now have to continue to turn the project into reality. Reviewing and monitoring the project is part and parcel of the whole process, and during the implementation process you may develop new approaches and have additional thoughts to be used during your next assignment.

Projects will inevitably take a substantial part of your working week. Attending or running meetings will also take a percentage of your time, so it is important to ensure that you are attending appropriate sessions, and if you run a meeting that you make the best use of the time.

5.3   Attending and running meetings

A meeting is an expensive use of an organisation's most valuable resource: the time of its employees. It follows that meetings should be carefully planned and executed efficiently. Meetings must have concrete objectives, usually agreeing actions or changes, and these should either be achieved, or conclusions drawn as to why not, and what has to be done next.

We are primarily concerned here with formal meetings, though informal ones, for example when a colleague or member of staff drops in to ask a question or seek guidance, benefit from similar planning. It is clearly better for the person seeking the meeting to say, 'Can you spare me ten minutes later in the day to discuss this problem?' rather than just barging in and expecting the other party immediately to redirect his or her thoughts to the matter.

5.3.1   Is your meeting necessary?

Think through all the meetings you attended in the last month and analyse the motivation for organising them:

- Because you always have one on the first Monday; or you set the date last time.
- Crisis management: one of your factories blew up last night.
- Someone wanted to pass on some good/bad news and see how you reacted, or wanted help.

Some of your meetings were probably necessary, some simply should not have happened and others might have been equally well managed using electronic mail/media, fax messages or telephone calls.

In future, before calling a meeting or agreeing to attend one, ask yourself, 'Is this really necessary?' and 'Is there a better way?'

5.3.2   The categories of meetings

Meetings not only take time but have underlying costs. There is the direct cost of bringing people together and holding the meeting itself, but also the lost time for those involved not doing their normal work.

Formal meetings fall into four general types:

- **Creative meetings**
  For finding new solutions or improving existing ones. Quality circles, brain-storming sessions and freewheeling are examples of creative meetings. Number of participants should be kept below about ten. The meeting convenor should prepare and circulate a list of objectives some days beforehand so that everyone has an opportunity to prepare.
- **Action and objectives meetings (or review meetings)**
  Assessments are made of projects or people and actions decided. Keep to less than fifteen participants; everyone must prepare their contributions in advance, keeping them relevant to the agenda.
- **Information and motivation meetings**
  Used primarily to pass on information, for example about new products, new offices or new working practices, to potentially large numbers of people. The agenda will list a number of speakers with pre-prepared presentations; the audience is usually passive.
- **Legal meetings**
  Called to satisfy legal or regulatory requirements, such as Board Meetings, Annual General Meetings and the like. The agenda lists the speakers and the atmosphere is formal, sometimes with pre-prepared contributions from the floor.

Of these four types, we are going to be focusing on the first two, where individual team leaders and other participants are able to set the agenda, the discussion and the results. Legal meetings are beyond the scope of this book, as are information and motivational meetings; the latter will often be accompanied by morale-boosting activities, such as away-day weekends.

5.3.3   Preparation for the meeting

The purpose of preparation is to ensure that all of those attending understand why they have been invited, what they are expected to contribute and what decisions are to be taken.

The agenda should be very narrowly defined, so that everyone invited can contribute to each item. However, sometimes a great deal of expense is involved in gathering the participants together, so a wide-ranging agenda may be necessary. Most project-oriented meetings are of the first kind, with a typical agenda boiling down to:

- Where are you compared with the plan, time and budget? (Have any items joined or left the critical path?)
- What unexpected successes have you achieved?
- What unexpected problems have turned up and what will you do about them?
- What are the plans for the next period?

A headquarters meeting of all managers/team leaders, on the other hand, may expect a wide-ranging agenda extending over two or more days, with subjects as diverse as financial achievements and plans, economic forecasts, progress reports relating to new or revised products, marketing and merchandising plans, new or revised personnel policies, information about joint ventures, mergers or acquisitions and a variety of other material. One reasonable justification for such a workload is that, having gone to the considerable trouble and expense of gathering everyone together, it makes better use of time for them to be given fully comprehensive information.

This brings us to the next point: who to invite? An old joke has it that the best meetings have an odd number of attendees (to ensure a clear decision), less than two!

Most of us have been to meetings where our presence was requested more from politeness than necessity. Our interests would have been equally well served by a copy of the agenda and working papers, maybe a short discussion about one or two items, and a copy of the results. So ask yourself, 'Is this person's presence really necessary?' Also ask yourself if failing to invite someone will cause grave offence. Part of the role of a team leader is to maintain morale, so avoid needlessly offending people.

It should be clear that writing the agenda and the list of invitees are two sides of the same coin. A very narrow, even one-item, agenda probably involves just a handful of people. The opposite is also true. So consider carefully whether or not it might be better to have several distinct meetings, perhaps run sequentially so that people can stay over from one session to the next.

The purpose of the meeting(s) is to make concrete decisions, so when planning your meeting(s) keep this in the front of your mind. Ask yourself, 'Is this the most effective way overall of reaching my objectives?'

A final element in this generalised part of preparation is to consider the timing of meetings. Sometimes you have no choice: there has been a disaster and your crisis management team must get together – the fact that it is Sunday morning or Christmas day is irrelevant, the factory is on fire or the office has been blown up, and there are things to do.

In more ordinary circumstances you want to make good use of time and not cause needless personal inconvenience to participants. If people have to travel long distances then near the middle of the day may be more appropriate than early morning or late afternoon. If you want a quick meeting and everyone normally works in the same building, try 4.45 p.m. on Friday! A long meeting might straddle lunch, so that attendees can stretch their legs, maybe look at their email, and catch up with visiting colleagues from other sites. Not a big deal, but worth a few minutes' thought.

5.3.4 Stage two of preparation

Let's look at a few simple 'mechanical' aspects of preparation.

Invitations should contain the names of all those invited, the venue (with a map and details of public transport or parking facilities as relevant), the date, start and end times, and the agenda with overall objectives and item-specific objectives. Anyone expected to make a presentation should be identified and all participants should be formally instructed to make relevant preparations.

Most people find it quicker and easier to read information than listen to other people reading it. So insist that all formal presentations are circulated to participants long enough beforehand for them to have an opportunity to digest the information. This element of preparation saves enormous amounts of meeting time.

Agenda items should be listed in order of priority. As we have mentioned above in the time management section, a prioritising process should be used along the lines of urgent/not urgent and important/not important. Put the urgent and important matters at the top of the agenda since participants will still be fresh (and it will encourage punctuality). If individuals are not needed for all items then juggle the list so that they can arrive late or leave early, rather than rushing in and out.

Don't have an 'Any other business' item. The objectives of the meeting are clearly set out, and if someone has something to contribute which is relevant to an agenda item, that is when they should speak. 'AOB' is an opportunity for waffle, for time-wasting and for the riding of irrelevant personal hobbyhorses. If someone wants to raise an additional topic relevant to either the main subject matter or to the people at the meeting, have it added to the agenda and so dealt with properly.

Other obvious preparatory measures include booking the room and making sure it stays booked (and that there are enough chairs). Organise coffee/tea/lunch as appropriate. Assign someone outside the meeting-room to take messages on behalf of those attending. Make sure any visual aids are present and working. If relevant, assign a note-taker.

One final point relating to preparation, if you are going to be late despite your best intentions – your train is running late or the motorway is blocked – it is only polite to try to get a message to the convenor before the starting time, so that if your presence is essential to an early agenda item it can be rescheduled (or perhaps someone can deputise for you).

5.3.5   *The meeting itself*

An effective meeting is crisp, clear and as brief as is possible while reaching concrete objectives. The onus is on the chairperson to make this happen. This could be you or a colleague. A poorly chaired meeting tends to ramble, to wander off track exploring byways and to waste time.

If you are chairing the meeting it is therefore essential to set the correct 'tone' right from the start. This means starting at the pre-advised time of commencement, not ten or fifteen minutes later because a couple of people are late. Bang on time, 'Good day, ladies and gentlemen. Thank you for coming in today to discuss the main subject matter. Our first item is X. You have all seen the papers.' If you are chairing the meeting, summarise the objectives and answer any questions raised by the papers – and off you go.

Keep the objectives in mind and 'Go for the close' on each topic as quickly as possible. Some people make little prompt cards for each item with two or three phrases to represent the main decisions to be taken or questions answered and use them to keep discussions on track. Make sure everyone understands and agrees with (or has been overruled but accepts) the conclusions, and that they are recorded accurately on a protocol sheet before moving to the next topic. This sheet should include what actions are to be taken, by whom and by when. Get the collected protocols photocopied and distributed before the meeting breaks up.

The agenda should ideally have a time limit for each subject. If you make up time this will allow a little latitude for particularly contentious items or, better still, you can all go back to work (or home) a few minutes earlier. Finish by the appointed time, even if this means dropping the lowest priority agenda items. People often have other appointments which cannot easily (or politely) be delayed, such as planes to catch or other meetings, and an overrun early in the morning can cause chaos for the rest of the day.

Sometimes there is a threat of overrun because it becomes evident during the meeting that an additional topic really needs to be considered. Ideally, put it on the agenda for the next meeting or convene a special meeting restricted to just the relevant people. At worst, adjourn the present meeting for twenty minutes to allow some preparation time and then restart. However, the primary purpose of distributing working papers ahead of the meeting is to ensure that this sort of problem doesn't arise, since the need for extra discussion should be foreseen.

In keeping with this air of efficiency and onwards momentum, do not tolerate:

- digressions – no-one will have prepared so they just waste time;
- chatter by people not fully engaged in the subject-matter under discussion. If they don't stop talking, as chairperson stop the meeting, leave a pause, and then say you cannot continue the meeting until everyone is fully focused;

- mobile telephones in the meeting room – if people are expecting calls have them give their contact details to the appointed message-taker who can relay information at the next break;
- people using the telephone in the meeting room (if any); have the switchboard direct incoming calls to the message-taker and use the call-redirection feature to reroute calls from within the building to the same person.

5.3.6   After the meeting

Get the minutes written up while they are still fresh in your mind. The protocols recording specific actions should be appended, and everything distributed within one or two days to participants, to those who were unable to attend and, possibly section by section, to others with specific interests in individual items. The meeting should have induced some working momentum on the subjects discussed. Receiving the minutes within twenty-four hours sends a positive message and reinforces that momentum. A delay can have the converse effect.

If the meeting has involved extra work or particularly onerous travel for some individuals, write a short note to these people to thank them for their time and trouble: it boosts morale and does your personal 'image' no harm.

If you called the meeting, then the work described in the protocol sheets is presumably necessary to enable you to meet your own objectives. So follow up. Mark in your diary the deadline dates, and reminders to yourself to check progress a few days before the deadlines (or earlier/more often as the case dictates).

5.4   Chapter summary

- Manage your time, and undertake tasks and activities that are appropriate. Use the delegation matrix to delegate where feasible.
- Project management is more than an operational activity. It is about getting people on your side, convincing them that you are a good project team to work with. This will result in others acting as an 'active reference' for you personally and for the project team you are involved with.
- Be sure you gather in-depth information about the brief by asking the right questions and actively listening.
- Make sure people attending the meeting understand why they have been invited, what they are expected to contribute and what decisions are to be taken.
- Keep the meeting on track, and create a protocol sheet for ongoing actions required and timeframes to be adhered to.

Chapter 6

Recruitment and induction

This chapter will review the key steps of the recruitment process, defining the job description and candidate requirements, assessment criteria and the interview itself. It will then move on to the induction process, examining a best-practice approach in order to ensure the new team member fits into your team and achieves maximum performance as soon as possible.

6.1 Recruitment error

Recruiting the right persons is one of the most important things a team leader does. What constitutes a true error in recruitment?

It is to discover faults, shortcomings and incompatibilities after recruitment when they could have been identified before.

The goal of recruitment is to realise the objectives of controlling change. I cannot emphasise this enough. If you recruit the wrong people into your team you will never realise your objectives or proactively control change.

Business is about people. Why do staff leave organisations? In our experience it's not that they leave the organisation but that they leave people. Recruitment techniques are not enough to guarantee success. Why? Not everything is objective! As preparation during the selection process most team leaders and managers define objective requirements: qualifications, experience, etc. However, the more candidates you have who satisfy the objective aspects, the more important subjective criteria becomes. Ultimately this can represent 90 per cent of all requirements in the final decision.

Online recruitment

Most employers and HR software suppliers agree that using only online technology is not the best way to recruit people. Key word recognition can work up to a point in terms of sifting qualifications and experience; however, it cannot screen qualities such as potential or ambition. The face-to-face interview is still a critical part of the process, particularly if you are recruiting a person to be part of a team.

However, the Aberdeen Group produced a report ('The global war for talent: getting what you want won't be easy'),[1] where it identified how 'best-in-class' organisations are integrating technology into their recruitment strategies by profiling what top talent looks like, then connecting those who match – active job seekers or not – through social networking sites, search engines and email.[2]

Organisations vary greatly with regard to their policy of hiring staff. Blue chip organisations such as P&G hire new staff right out of school or university, and develop them through progressive work assignments and training. It would be unusual for P&G to bring in external talent other than at a junior level, except in the rare case of an acquisition like Gillette. It is almost unheard of for them to bring outsiders into a senior executive position. There are organisations, however, that choose to 'buy in' expertise, because they have a unique skill set for a particular job, or have the capability to attract a certain type of business.

Obviously, the purpose of recruitment is to find the best person for the job, but you need to attract them to apply in the first place.

Many organisations use recruitment to build a positive image of their organisation among all candidates who apply. Bear in mind that attracting and then keeping good people is vital for budgetary control purposes. Constant re-recruiting is an expensive operation, not just in the recruitment process itself but also in the reduced productivity levels of new hires during the induction process.

So, promote your company in the most positive way. If you look in the marketplace, an organisation such as Exxon Mobil is highly regarded within the energy industry for its training and development policies. Ketchum, the global public relations firm, also pride themselves at providing the best learning opportunities in the business. Most applicants are looking for interesting work, an appropriate remuneration package and people they feel they can get on with. They also want to know that there will be an opportunity to develop their skill sets in some way, so emphasis what form of training and development might be on offer to your candidate.

What other benefits can you highlight? There are obviously a range of bonus/ financial incentive/pension schemes that you might be able to offer, but what other forms of incentives do organisations offer?

Benefits case studies

A London-based building and construction industry recruitment company, the Consensus Organisation, put a healthcare strategy in place when the firm was launched in 2001. They produced an online health and wellbeing service, offering advice on nutrition, sleeping and exercise. They offer free fruit smoothies and there is a part-company-paid scheme for membership of the nearby gym.[3]

GlaxoSmithKline (GSK) offers financial education, a free consultation with an independent advisor, monthly financial surgeries, lunchtime presentations on financial issues, pre-retirement workshops and online webinars.[4]

Blue chips such as Unilever provide some of their product range at a substantially reduced price to staff. They are also known to provide free ice cream from a freezer in their office.

In the financial sector, most of the players provide subsidised loans to their staff.

Check with your HR department to find out what is on offer. If you work in an SME and have less formal processes, procedures and benefits policy discuss with your manager what benefits might incentivise the high-calibre applicants you wish to attract.

The candidate you interview for a particular job may not be suitable now, but what about the future? Consider all options before you decide to hire:

- Do you still need the role? If so, has it changed or could you reorganise it or redistribute responsibilities?
- Is there someone in your team who could take on more responsibility or additional challenges?

Once again, ask for help from your HR department if you have one. Work with them, use their expertise and save yourself time!

6.2   Defining the job and candidate requirements

Things to consider during the interview preparation stage:

- Start by thinking about key skills, competencies and other considerations that you are looking for in the CV. This will allow you to properly screen CVs and prepare the appropriate questions to ask during any interview session.
- Use the job and candidate requirements to prepare specific questions to ask an applicant if there have been any gaps identified when screening his/her CV.
- Remember that qualifications, education, country experience and languages are *objective*. The more candidates that satisfy these, the more important it is to uncover additional evidence based on your prepared questions.

Identifying job and candidate requirements serves as a guide for scoring/evaluating CVs. This is the primary instrument of recruitment. The more concrete its definition, the better your time and energy will be applied in the selection process.

- Define the role and job description.
- Define the job requirements.
- Define the candidate skills, competencies and attributes.
- Determine screening and assessing criteria.

6.2.1   Define the job description

The job description helps you to understand exactly why you are recruiting and what you are looking for. As the hiring team leader you have the authority to amend job descriptions to reflect the actual role. How you define the job is your decision. Should you amend a job description, discuss this with your manager/HR client manager.

The more concrete the job description the easier it will be to find a good candidate. Defining the job description is not simply a bureaucratic exercise; it is an essential part of recruitment. This responsibility is the hiring manager's and not HR's.

6.2.2   Define the job requirements

Ensure the essential job requirements are incorporated in the job description.

- job description responsibilities and deliverables
- education and qualifications
- experiences/country experience
- languages
- ongoing activities or forthcoming special projects you are aware of for this position

6.2.3   Define the candidate's skills, competencies and attributes

Make sure the essential candidate skills, competencies and attributes for this job are incorporated in the job description.

- A *skill* is the capacity to successfully accomplish something requiring special knowledge or ability. Examples include: type 70 wpm or speak Russian.
- A *competence* can be described as 'an underlying characteristic of a person which results in effective and/or superior performance in a role'. Examples include: achieve results or provide direction.
- An *attribute* is a quality or characteristic that a person is born with. Examples include: highly-strung or tolerant.

6.2.4   Determine screening and assessing criteria

The criteria used to screen and accept or reject CVs are based on the job and candidate requirements, listed in the job description under 'Essential skills, experience and qualifications' and 'Competencies and personal attributes'.

- **Determine screening criteria** – What are *essential* job and candidate requirements I must identify to screen CVs?
- **Determine assessing criteria** – What questions must I prepare to probe and clarify whether essential job and candidate requirements mentioned in the CV are valid?

If a candidate does not possess all the screening criteria *but* is working to achieve these (for example, has not yet completed a particular qualification) and *seems* to possess the competencies and personal attributes required (team player/self-motivated, etc.) you may decide to interview them.

It is best to involve all stakeholders during the interview process, so it is advisable to prepare a panel. You should create a list of agreed questions that you will ask in turn. Each candidate needs to have the same questions asked to them for equal opportunities purposes. Probably around five questions is sufficient, and make sure the questions are open – in other words that they cannot be answered just by a 'yes' or 'no'. The interview session would be a little short otherwise! You should use general questions and specific biographical and competency questions related to your team's key results areas (KRAs). You must think through how each question is formulated, and understand the results you are looking to achieve. Talk with your HR department to clarify what questions you must *not* ask to stay within legal, diversity or ethical best-practice norms.

6.3   Carrying out the interview itself

Since you want to optimise the time spent interviewing, you should aim to relax the candidate as soon as possible so that you can get them to 'open up' and explore factors that are important to the job.

Get the environment right. If it is going to be a one-to-one interview in your office, ensure privacy and no interruptions. Clear your desk as much as possible, keeping just the documents you require for the interview. This shows you are prepared and organised, and in turn respect the situation and the person you are interviewing. A desk will be a barrier, so arrange seating in an informal and relaxed way. If the interview session is to be with a panel, a meeting room would be best, ideally with a round table.

Use the interview to build rapport and create a positive image of your team and the organisation. The first minutes are crucial in setting the tone of the interview. So let's look at the interview in overview terms to start off with.

A good way to open the interview is to ask the candidate, 'What do you know about our organisation?' It is surprising how many candidates don't even bother to check the organisation's web page for basic information about the company. It is not a very promising sign if they have not done at least this basic research prior to coming to the interview.

You can then progress to laying out clearly and precisely the big picture information about the organisation (not too much detail here – the candidate does not need to know about every department/product range. You want the time to be spent with them talking, not you). Then quickly move on to talk about your team and the role that the applicant has applied for.

Use 'How' and 'What' questions wherever possible, and minimise the use of 'Why', as this can put pressure on the applicant; it suggests justification or defence is required. Since you want them to be relaxed to start with, if you need to use 'Why', it is best towards the end of any interview.

If a psychometric instrument has been used prior to the face-to-face interview, you will need to give feedback from the results. Position the test as a helpful guide for all parties, not the deciding factor.

Throughout the interview use open questions as clarified below.

Control the interview by active listening. Take notes and ask probing questions that help you to concretely understand what exactly the candidate did in previous roles.

Time should always be left for the candidate to ask questions themselves. This is where you get a real feeling about them as a person. What is important to them? How much do they understand about the roles and responsibilities of the post they are applying for?

End the interview positively by explaining the next steps and what will happen and by when.

6.3.1   *Various interview questioning techniques*

Open questions

Use 'open' (who, what, where, why, when, how, which) questions to open up the discussion. Open questions allow candidates to describe events from their own point of view. Examples might include:

- 'What did you like best about your last position?'
- 'How do you see your future developing?'
- 'What attracted you to apply for this post?'

Closed questions

Closed questions start with a verb, 'did you ... ?', 'have you ... ?', 'will you ... ?', and only require yes/no answers. These tend to close down conversation. Closed questions can be used to check facts or verify answers. Examples might include:

- 'Did you succeed in doing that?'
- 'Did you say that you worked in a development department for two years?'
- 'Have you any more information to add?'
- 'Do you consider the projects you were involved in too regional?'

Biographical questions

Biographical questions relate to *past* jobs, life experiences and education that have helped the candidate to excel (competence) in their career. Examples might include:

- 'What have you learned about your strengths as a result of working at the various jobs you have held?'

- 'What do you think of your career progress so far?'
- 'How do you think your work experience and education will assist in this role?'

The benefit to interviewers of using these questions is that they reveal life and relevant work experiences and education required in various roles. These questions are used to get candidates to describe, in detail, the relevance of past positions to future roles and their responsibilities.

Competency questions

Competency questions describe 'how well a job is performed' or are used to describe what excellent performance should look like in a job. A competency is a fairly deep and enduring part of a person's personality and can predict behaviours and perform-ance required to successfully meet the job tasks and responsibilities. These questions are used to get candidates to describe in detail the behaviours, skills and attributes required to meet job tasks and responsibilities in the role you have in mind. Examples might include:

- 'How do you make sure the services you deliver meet customer needs fully?'
- 'Tell me about a demanding goal you set for yourself'.
- 'Tell me about a time when you have really had to dig deep to find out about an underlying problem'.
- 'How do you behave under pressure?'

Probing questions

Remember to follow up on answers given. While the wording of the opening question is vital, follow-up questions are essential to allow you to *probe* deeper. Probe until it is clear what the candidate actually did. Never allow candidates to get away with gener-alisations. This is one of the most frequent errors made in interviews. Probing for full and accurate information requires skill and requires the interviewer's perseverance if certain areas of information are being avoided by the interviewee. Examples might include:

- 'Tell me more about it . . . '
- 'Can you think of another example where . . . ?'
- 'What happened . . . ? What was the outcome . . . ?'
- 'What did you actually do or say?'

Leading questions

Avoid leading questions. These types of questions imply a particular answer is correct and so lead the interviewee to give the reply s/he thinks you want to hear. An example of a poor leading question might be:

- 'We seek proactive people. How would you describe yourself – proactive or reactive?'

Double or multiple questions

Avoid double or multiple questions. With multiple questions, there is a tendency for the interviewee to only answer one part of the question, for example:

- 'What is it that you enjoy most when working with people – having staff to manage or the challenge of motivating them?'

It would be better to ask:

- 'What do you enjoy most when working with people?'

Theoretical questions

Avoid theoretical questions. You should be more interested in knowing what a candidate would do when faced with a real situation, rather than if they know what they 'should' do. Therefore you want to seek out examples of what the interviewee has actually done when faced with that situation, for example:

- 'Sometimes we give feedback and it is taken in the wrong way. What would you do if this were to happen to you?'

It would be better to ask:

- 'Sometimes we give feedback and it is taken in the wrong way. What have you done in the past to ensure that your feedback is understood in the manner you meant it?'

Reliability of evidence

Encourage and facilitate potentially negative answers. Phrase questions to enable potentially negative answers to be given more readily, for example:

- 'We all have to lower our personal standards sometimes. Can you tell me when you have had to compromise in this way?'

Ask questions that cover several situations to build up reliability, such as:

- 'We all have to lower our personal standards sometime. Can you tell me when you have had to compromise in this way in the office and with a client?'

Asking questions related to the candidate's CV

One section in a CV can be a hive of information. For example under 'sample achievements' a candidate's CV states, 'I managed my team's capability development to embed the new OD and business processes, which raised the professional standing of the department with our customers. This included team building, through coaching, support and the use of challenge techniques'. Related questions you might ask here are:

- 'How did you actually manage this?'
- 'How many people were in your team?'
- 'What did "capability development" actually mean in this instance?'

- 'You say that you raised the team's professional standing. How did you measure this?'
- 'Give me three examples of "challenge techniques" you used.'

Questions you should NEVER ask

In every job interview, the goal is to obtain important information while building a friendly rapport with the candidate. But some questions are just a little *too* risky. Protect yourself and your organisation from legal trouble and embarrassment by avoiding the wrong questions while still getting to the root of the concern behind the question. Do not discriminate. You must *never* ask questions – this is a legal requirement – which might be thought to discriminate against a person's sex, sexual orientation, age, race, religion, marital status, pregnancy, disability, political views, etc. Also, stay away from making personal comments or statements during interviews. Candidates know their rights and will judge the interviewer and your company.

Looking at the following sample interview questions you can see which ones are acceptable or unacceptable to ask. Just be careful how you ask your interview questions. You must know how to turn possibly litigious questions into harmless, legal alternatives.

Which of the following are acceptable or unacceptable?		Acceptable	Unacceptable
1	Are you an EU citizen?		X
2	Are you authorised to work in the UK?	X	
3	Which religious holidays do you observe?		X
4	Are you always able to work within our required working week?	X	
5	Do you belong to a club or social organisation?		X
6	Are you a member of a professional or trade group that is relevant to our industry?	X	
7	How old are you?		X
8	Are you over the age of 18?	X	
9	Do you have or plan to have children?		X
10	Can you get a babysitter on short notice for overtime or travel?	X	
11	What do you think of inter-office relationships?		X
12	Have you ever been disciplined for your behaviour at work?	X	

Why are the above examples unacceptable?

1. Although this seems like the simplest and most direct way to find out if an interviewee is legally able to work for your organisation, it's hands-off. Rather than enquiring about citizenship, question whether or not the candidate is authorised for work.

3. You may want to know about religious practices to find out about weekend or other work schedules, but it's imperative that you refrain from asking directly about a candidate's beliefs. Instead, just ask directly when they're able to work, and there will be no confusion.

5. This question is too revealing of political and religious affiliations. Candidates are not required to share such information with potential employers. Additionally, this question has little or no relation to a candidate's ability to do a job. Ensure that your wording focuses on work.

7. While maturity is essential for most positions, it is important that you don't make assumptions about a candidate's maturity based on age. Alternately, you have to be careful about discrimination towards applicants nearing retirement. Knowledge of an applicant's age can set you up for discrimination troubles down the road. To be safe, just ensure that the candidate is legally old enough to work for your firm.

9. Clearly, the concern here is that family obligations will get in the way of work hours. Instead of asking about or making assumptions on family situations, get to the root of the issue by asking directly about the candidate's availability.

11. The practice of inter-office relationships can be distracting, break up teams and cause a number of other problems in the workplace. But asking this question makes assumptions about the candidate's marital status and may even be interpreted as a come-on.

Speak with your HR department to find out more about the questions you should not ask. Keep abreast of changes in legislation.

Check hobbies

You can identify what is really important to the candidate by knowing what *motivates* him or her. People's motivations relate to their values. If you can discover these it may allow you to find out what candidates may be prepared to commit to.

Example: Sailing
'What are you looking for when you go sailing?'
Answer 1: Getting away, being alone, thinking, forgetting, planning...
Answer 2: Finding a common focus of interest with friends, teamwork, forgetting the hierarchy ...

Example: Free time
'What do you do in your free time?'
Answer 1: Spend time with my family, watch TV ...
Answer 2: Meet friends, manage a sports team ...

Summary of the top 11 recruitment pitfalls

1. Failing to look in-house.
2. Looking for an exact personal replica 'mirror' or superhero.
3. Not explaining the process or next steps and when they will happen.
4. Going it alone – not involving HR or key stakeholders.
5. 'Winging it' that leads to a poor interview structure and using the wrong questions.
6. Going with 'gut feel' or 'first impressions'.
7. Missing signals.
8. The candidate not talking. The candidate should talk for 70 per cent of the interview at least.
9. Poor follow-up/selection process takes too long.
10. Over-promising, under-delivering.
11. Lack of professionalism.

6.3.2 Hard evidence

You will need to write up hard evidence you observed during the interview and give concrete written feedback to HR/others to allow them to inform candidates of how well they did. Candidates often request feedback, particularly if they work in-house already.

Finally, the costs of one recruitment error can be enormous. We always ask ourselves 'would I enjoy working with this person?' and 'what value will this person add to the business?' If there's any doubt, we always say 'no'.

6.4 Induction

When we go into organisations to do research we find that the induction process for new hires can often be quite skimpy. The aim of induction is to help new employees make a smooth, positive adjustment to your workplace. Induction must be designed to assist team leaders in providing new employees with valuable information about their organisation and their new job.

The aims of induction are to:

- support your organisation's business objectives through swift and effective introduction of a team member into their new position
- reduce the anxiety often experienced by changing job/organisation
- foster a positive attitude towards your organisation
- answer questions not handled at the time of recruitment
- reinforce or establish realistic job expectations
- through the provision of information and a well-planned induction programme, increase knowledge of the organisation and the job, so developing the confidence and motivation of all new staff

6.4.1   What are your responsibilities for induction, as team leader?

- Provide the new team member with a positive role model.
- Provide specific information about the team and department structure:
  - organisation and structure
  - operational activities
  - relationship of function to other departments
  - goals and current priorities
  - introduce the new person to the team and job
  - detailed explanation of job description
- Coordinate and individualise the induction activities.
- Reinforce the new team member's introduction to your organisation.
- Make sure you are available and accessible during the first few days and weeks.

6.4.2   Why is proper induction so important?

First impressions last

When a new person joins the team, it is very important that they have a good impression of the organisation itself and of you as the team leader, and see how professionally and efficiently the team itself operates, as this will affect their attitude to working with you and the team.

They will also need practical tools and equipment to do their job, so that they are able to make a positive contribution to your team immediately. Saying this might seem glaringly obvious, but we know from experience that a new person can often be squeezed into a space which is inappropriate for their requirements. We have seen sales staff who are on the telephone all day being put next to photocopying machines, a new hire being put into an enclosed storeroom area with no natural light and, of course, the three-legged table, with the fourth leg being propped up with phone directories!

So, in practical terms what should you consider?

Prior to a new team member starting, and their first day

1. Organise desk/office and appropriate equipment and supplies:
   - mobile telephone
   - email access
   - access to your organisation's intranet and websites
   - business cards
   - laptop and appropriate software
2. Prepare (with relevant personnel) their personal induction pack or employee handbook that includes a copy of their job description and all contractual terms and conditions.

3. Agree with your HR or relevant personnel, within the first three days of the new team member joining your organisation, when the new person should meet with them to go through your organisation's employment practices and benefits.

4. Arrange a mentor or 'buddy' to guide him/her through the first weeks in the organisation and to answer any questions they may have. The 'buddy' can give a tour of the organisation's facilities to include:

- rest rooms
- telephones/message systems
- mail room
- copy/fax machines
- conference rooms
- computers/printers
- parking

- equipment/supplies
- storage/files
- books/reference
- kitchen and bulletin board
- shopping facilities
- systems support
- emergency exits

5. End the new person's first day with a personal wrap-up to answer questions and make them feel at ease. At this time, it's a good idea to agree a date for an initial welcome lunch within the first two weeks.

Day two or three

Once the new hire successfully ends his/her first day, your responsibility for providing job orientation isn't over – in fact it has just begun. Early in the first week you will need to devise a specific orientation plan. This induction plan is one that you work out directly with the person concerned to meet his/her individual needs. To help you devise an effective, individualised induction plan consider these points and questions:

1. Early on during the first week, most likely the morning of Day 2, sit down with the new team member for about one hour to plan and agree with them the activities and goals for that week.

2. Does the new person understand the team's goals as well as their specific job details?

3. What does he/she need to know by the end of the first week?

4. What experiences from this person's first day need reinforcement?

5. Recall the wrap-up at the end of the first day and think about what the new team member may not have absorbed or had concerns about.

6. What key policies and procedures do you need to convey during the first week? Those items critical to this person's job success are best handled now.

7. What positive behaviours do you want to reinforce during this first week?

8. What should you do to help integrate this person into his/her particular work group?

9. How can you give this new person a sense of accomplishment during the first week? Prepare a list of specific work assignments that offer a rewarding experience for them.

10. What feedback will they need from you?
11. How can you make yourself accessible? The hours you are able to invest during this first week may save you hundreds of hours in the months to come. Assess what you need to do with the new team member, estimate how long it will take, schedule the activity and then commit yourself to carrying it out.
12. Plan to meet again at the end of the week or the start of week two for a review meeting. This review meeting enables you and the new team member to review the orientation plan and check off the items achieved. This process visibly confirms the person's accomplishments and progress. During the first month of employment, meet with him/her on a weekly basis.

As a help, here is an example of a protocol sheet that you could expand or develop for your particular requirements, in order to monitor progress of the new team member:

Session:	Persons present:			
Date:	**Copy to:**			
Decision	*Who?*	*By when?*	*Time*	*Done?*
Go over team/dept. and organisation charts and identify the new hire's role	Team leader		1 hour	
Attend team staff meeting	New employee		2 hours	

After one month in the organisation

View a new team member's orientation and training as a continuous process rather than as a single event. Many induction programmes we have been involved with lasted up to one year to complete. Your task as team leader is to provide all the information and tools they need to work effectively and productively.

Induction questions for team leaders to ask themselves after one month

Before the end of their first month, provide additional details about their role in the immediate work group, department and division. In addition, be sure to provide the new team member with more information about their specific job. The following list of questions will help you clarify what should go into the evolving orientation plan for a new employee.

1. What additional things does he/she need to know by the end of the first month? Review your earlier lists for items that might extend into a long-term process.

2. What policies and procedures could affect this person's job performance? Does he/she understand all policies and procedures?
3. What impressions or values do you want to reinforce to him/her?
4. What specific tasks can you assign to him/her that will allow for growth? Examine tasks performed by people in the immediate work group, and structure the simpler ones into interesting and challenging assignments for the new person.
5. What can you do to broaden your delegation of authority and decision making? Concentrate on getting specific tasks done and increasingly delegate to the new team member as appropriate.
6. What training objectives do you want to meet within the team member's first three to six months?
7. Hold a review session after one month. See example induction questionnaire below.

Name: _____

Please answer the following questions in writing and bring them with you to our meeting, scheduled for __/__/____

1. What were your ambitions when you started with us and how, if at all, have these changed?
2. What have you found satisfying about your work in this first month?
3. Is the work you are doing in line with the job description and position you were recruited for?
4. What kind of difficulty, if any, have you had in positioning yourself in the organisation?
5. What concrete assistance should I and the company give you in the next month(s) to make you still more effective?
6. What are you going to do to increase your professional efficiency in the next month(s)? Concrete points.
7. What are your suggestions for improving the way we induct new employees?

When your team member has been six months in the organisation

The following questions can help you, the team leader, to check if the first six months' orientation and training of your new team member was effective. These questions are just for you and can help you ensure that any future new team members get the best possible welcome into your team.

1. What additional information would have helped my new team member during their first six months? Refer to your organisation's employee handbook for

relevant items not previously covered. For example, they may need to know about safeguarding confidential information.

2.   What additional policies and procedures does your team member need to better understand or have reinforced?

3.   What more can you do to reduce the time needed to manage this person? There are many learning experiences such as working with a more experienced member of the team that might help.

4.   How can you broaden your new team member's assignments so that they are continually challenged? The team member will experience growth in performing their job, and you need to match that growth with broader assignments.

5.   What feedback should you give on your new team member's performance? If you meet regularly with him/her and monitor the progress of their orientation plan, you may already be providing some feedback on their performance. You may also consider providing a performance appraisal at the end of the person's first six months.

6.   What training objectives do you want the new team member to meet within his/her next six months?

Feedback

During the induction process there should be opportunities for the new team member to give you feedback. It is not a controlling activity from you as a team leader. So here are a few ideas of questions you might ask the new hire. Give them the questions in advance of a meeting with you to give them time to prepare.

- Generally speaking, what do you think of your recruitment process?
- Which of our team's/organisation's strong points struck you first?
- What was the first fault or weakness that became apparent to you?
- What reassured you?
- What worried you?
- Looking back, what useful information was not communicated to you when you were hired or started the job?
- If you were to employ someone into the company, what would you do in addition to what you yourself experienced?

6.5   Chapter summary

- A new hire is an important decision to make. Put time, thought, and a structured approach into the interview process, to make sure you can find out as much as you can in a relatively short period of time
- Think about the organic growth of the team – is the new hire right for now, and also for future requirements?

- You need to give the appropriate support to him/her through a robust induction process. Openly encourage feedback through a strengths and weaknesses questionnaire at the end of six months
- Above all else, make them feel welcomed, that you are there to help, and ensure they develop an understanding of the values and culture prevalent in the team. They also need to be clear about the performance criteria against which they will be assessed.

Chapter 7
The art of motivation

Your behaviour as a team leader has a direct impact on your team members' performance, productivity, satisfaction and turnover. In this chapter, we examine the qualities of team leaders who motivate, outlining some proven techniques to inspire those who work for you. We also consider here the team leader's individual attributes and competencies necessary to inspire the right motivation for team members to undertake a particular task or workload.

7.1   Why is motivation important?

We all recognise the importance of motivation, whether in our personal lives or in the workplace. Highly motivated individuals and teams have a 'buzz' which enables them to move mountains. Weakly motivated people give up at the first hurdle; they simply do not have the will to rise to the challenge. In extreme cases the outcome is literally life or death, as when highly motivated soldiers, long-distance sailors or survivors of remote shipwrecks or plane crashes come back, while their less motivated brethren disappear without trace.

Perhaps the single most important technique for motivating the people you supervise is to treat them in the same way you wish to be treated: as responsible professionals. It sounds simple; just strike the right balance of respect, dignity, fairness, incentive and guidance, and you will create a motivated, productive, satisfying and secure work environment.

Unfortunately, as soon as the complexities of our evolving work life mixes with human relationships, even the best-intentioned team leaders can find the management side of their jobs deteriorating into chaos. Today's engineers face expanding workloads, fewer resources, greater customer expectations, increasing threats (e.g. malpractice lawsuits) and closer scrutiny, especially from customers and third-party providers. The art of engineering is being transformed into a business. And like it or not engineering practitioners find themselves in team leadership roles, with tremendous responsibility and, sometimes, little real authority.

Job performance is reflected more in the bottom line than in the quality of team leadership. Why, in this environment, do some team leaders thrive while others burn out? The answers lie in each manager's ability to inspire trust, loyalty, commitment and collegiality among team members. The same techniques that work elsewhere in business can bring success in engineering – whether you're working in an engineering practice, administration or academia. More often than not, though, the task can be accomplished only by replacing learned behaviours with newer, more effective models.

7.2   Unlearning autocratic style

Good team leadership technique used to be simple. The team leader told employees what to do, and they complied. No-one worried if anyone's feelings were hurt along the way. Team members who failed to toe the line were either whipped into shape or fired. These authoritarian team leaders believed that authority should (in a moral sense) be obeyed. Therefore, they expected unquestioning obedience from their subordinates and they, in turn, submissively obeyed their own superiors. What could be simpler? Fear ran the work setting. The system was efficient.

The example set by past generations has led to huge numbers of autocratic managers today. Some lead this way because they honestly and consciously believe it is the best management style. For most, however, it is how they were treated throughout their careers (particularly in a first job). The cycle works very much like child abuse, where the abused child grows up to be an abusive adult. If you were managed by an autocrat, it is very likely that your most natural, comfortable method of management reflects that of a previous manager.

7.2.1   Why change?

While fear as a management style can accomplish impressive short-term results, the long-term consequences can be devastating. With the world economy the way it is, no engineer or team leader can afford to alienate anyone. Similarly, efficient team members are also becoming harder to recruit and train, as the technology of the workplace speeds along at a blinding pace. Disgruntled team members may vent their frustrations by being rude, performing poorly, quitting or complaining to upper management; some team leaders may even face lawsuits for treating subordinates unfairly.

An autocratic management style feeds high staff turnover and low employee morale. Low morale, in turn, causes a decline in productivity and in the quality of service provided to your customers. And while many autocratic team leaders still populate the workplace, reform demanding higher efficiency and productivity will eventually squeeze such team leaders out. In short, motivational management produces better results; those who focus on positive reinforcement rather than fear and intimidation will be the successful team leaders.

7.2.2   Movere: motivation = setting in motion

Motivation comes from the Latin word 'movere' which means 'to move'. Motivation therefore means 'setting in motion'. Umberto Eco wrote, 'today, "movement" is so swift that science fiction novels become archaeological treatises as soon as they are published'.

The level of motivation of team members can increase productivity and quality levels by up to a factor of two. The impact of this difference on competitiveness is hard to over-state, and high levels of motivation maintained consistently over the long term can easily make the difference between business failure and success.

What are the things that really motivate team members? When we deliver a training session we sometimes ask the following questions. (By the way, how would you answer them?)

1. For what two specific reasons is it important to know your team members' strong points?
2. What are the three most productive forms of motivation?
3. What, in your view, are the three most important specific qualities which determine our success as managers?

And the *correct* answers are!

1. To make them the best in their field! Assign tasks and set targets on the basis of your team members' strong points. Do you really know the strong points of your team members? I urge you to ask your closest team members to write down what they think are their three strongest points and compare your answers with theirs. This can make for interesting reading!
2. Ask them 'what do you propose?', praise them when praise is due and set concrete objectives.
3. Listening, controlling change and being a positive role model.

Exercise

**Identify three motivational leaders or managers in your organisation.
List the qualities and behaviours that make each individual a motivator.**

1. Name _____

Qualities	Behaviours

2. Name _____

Qualities	Behaviours

3. Name _____

Qualities	Behaviours

Typically, a leader who motivate others demonstrates some of the following attributes and behaviours.

1. Focuses on self-improvement and development as a leader.
2. Teaches and coaches.
3. Helps individuals recognise their potential and assists in the development process as a coach and mentor.
4. Provides frequent, open and honest feedback that reinforces positive behaviours and critiques areas for improvement.
5. Brings out the best in others.
6. Is recognised as being trustworthy and honest; others seek this person out as a mentor and confidant.
7. Encourages risk taking.
8. Celebrates others' successes and contributes to building others' self-confidence.

Using a scale of 1–10 rate yourself on each of the above points.

To score a ten (the best score) you must demonstrate this skill 100 per cent of the time.

Attribute	Score
1. Focuses on self-improvement and development as a leader.	——
2. Teaches and coaches.	——
3. Helps individuals recognise their potential and assists in the development process as a coach and mentor.	——
4. Provides frequent, open and honest feedback that reinforces positive behaviours and critiques areas for improvement.	——
5. Brings out the best in others.	——
6. Is recognised as being trustworthy and honest; others seek this person out as a mentor and confidant.	——
7. Encourages risk taking.	——
8. Celebrates others' successes, and contributes to building others' self-confidence.	——

If you did not rate yourself a 10, identify one way you could raise your level of expertise in this area. Be specific about what your action is. Consider what the person you feel does motivate might do to improve. If you did rate yourself a ten, identify three ways you increased your expertise in this area.

It is common to find that a highly motivated workforce runs rings round a demotivated team, and industrial psychologists have devised experiments to quantify the difference which, in slightly exaggerated form, may be summarised as 'two for the price of one'. High levels of motivation maintained consistently over the long term can easily make the difference between failure and success, growth and stagnation, becoming a recognised 'winner' and disappearing.

Some companies indicate their recognition of this by developing tag lines such as 'our people are our most important asset', but how many organisations truly and seriously make motivation a primary management tool? Do you, as a team leader, consciously mould each of your actions and decisions in the light of their impact on the motivation of your team? When was the last time a manager was removed for allowing the motivation of their team to drop below acceptable levels? What is the true state of affairs in your organisation?

- An occasional 'rah-rah' meeting supposedly designed to raise morale, but which in fact merely reinforces cynicism?
- Regular training in motivational techniques, or a limp statement of objectives with little or no follow-up?
- Tough appraisals of team leaders and managers, which make a central feature of their motivational skills and the level of motivation of their teams, with retraining or redeployment an immediate outcome of poor marks?
- Ongoing monitoring of morale right across the workforce, with reinforcement for high-morale teams and remedial treatment for the others?

We take it for granted that you and your organisation recognise the value of high motivation, wish to improve on your current levels and then maintain that improvement. But it is often difficult to know where to start. In addition, it is obvious that, just as you find that some motivating techniques work for you (and you find others childish and demeaning), the same will go for your team. They are individuals and respond in different ways to the stimuli you can give. So you need a whole range of techniques, and you need to learn how to apply them for best results.

7.3　First principles

'Motive' is an ancient and unusual word, in that it may be used as a noun, adjective or (rarely) a verb. Their meanings revolve around making something happen, initiating movement or motion. 'Motivate' is a relative newcomer to the family (1863), meaning 'to furnish with a motive; to cause someone to act in a particular way' (*Shorter Oxford English Dictionary*). So by your words and actions you can 'cause people to act' positively and productively, or negatively and destructively.

Can you predict the consequences of your causes? Yes, of course. The simplest means is to think about how you would react if you were on the receiving end, though bear in mind that different personalities react in different ways. A quick review of theory might also be useful. An American psychologist, A.H. Maslow, drew a 'pyramid of needs',[1] split into five bands which build up from the basic needs for food and shelter to 'personal fulfilment', the icing on the cake (Figure 7.1). This analysis is important for several reasons:

- It is true of all normal human beings, whatever their personality type or personal situation.
- The successive levels are arranged in priority sequence, so if you try to satisfy a high-level need before a lower-level need you are building an unstable situation (all the personal fulfilment in the world will not raise motivation if the individual isn't paid enough to provide adequate food, clothes and housing).

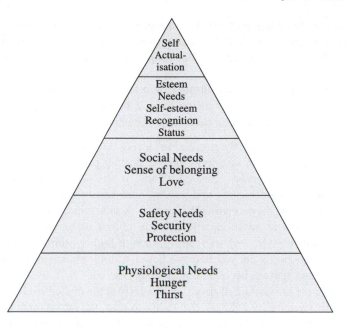

Figure 7.1   The Maslow pyramid, associated with Maslow's theory of the hierarchy of needs

- By understanding the needs, you can construct reward packages, including non-salary benefits and intangible benefits, *tailored* to the individual whose motivation you wish to raise.
- It provides a quick checklist which you can use to screen out potentially counter-productive initiatives (those which cause deprivation at lower levels).

The five levels are:

1. **Physiological needs:** basic food, clothing, shelter, etc. for the individual and his or her dependants, normally catered for through the basic salary.
2. **Safety needs:** ranging from physical security to long-term financial security; no-one can be highly motivated if their concentration and commitment are distracted by chronic fear. In today's world many aspects of this level of security are provided by insurance, pension schemes and the like.
3. **Social needs or belonging:** most of us need to feel accepted as full and equal members of the human race; we satisfy much of this need through our families and networks of relationships reflecting our particular interests, religion, politics and culture but, given that we spend around a third of our adult lives at work, a sense of 'belonging' to work-related groups has an important bearing on our attitude to our colleagues and employers.
4. **Esteem:** by and large we try reasonably hard to do a good job, and take pride in the fruits of our labours; we take even more pride and pleasure in seeing these

appreciated by others, particularly our superiors. The esteem of others raises our image amongst our peers, differentiates us within the group while reinforcing our sense of belonging and value to the group. It also acts as a spur to greater efforts in future.

5. **Self-actualisation or self-fulfilment:** that feeling of pleasure from 'a job well done', whether it is a neatly mown lawn, a beautifully designed power station or an ordinary repetitive task performed efficiently and effectively.

As you have already seen, the needs at the top of this list take priority. These must be satisfied before you can take more than a fleeting pleasure from having lower-priority, higher-'quality' needs satisfied. (There is obviously a need to keep these needs satisfied, but regular salary and benefit increments should do this.) Once they are satisfied, however, they are virtually ignored. A salary adequate to keep ourselves and our dependants in an acceptable lifestyle allows us to start worrying about maintaining that lifestyle, and so security takes a higher profile. Once these two 'inwards-looking' needs are satisfied we begin to look around ourselves and our place in the group starts to become important, and so on.

Meanwhile the satisfaction of needs, particularly at the higher levels, has an iterative element. When we are toddlers, the ability to read 'the cat sat on the mat' brings great satisfaction; a year or two later we need to read a complete book to get the same thrill. To borrow a phrase from the field of athletics, we gain self-fulfilment by constantly achieving new 'personal bests', stretching our talents to meet ever more ambitious goals.

7.3.1   Maslow in the workplace

It is nowadays widely held that the need for self-fulfilment and the need for esteem or recognition are amongst the most important motivating factors in the modern workplace. However, there is less agreement as to whether the sources of the satisfaction are intrinsic within the actual work itself or whether they lie in the environment, the quality of relationships between managers and employees and the atmosphere within the organisation.

If you think about this for a moment, it soon becomes clear that these are two sides of a single coin. No matter how satisfying your work, if it is conducted in a sour atmosphere you soon lose your will to excel. Similarly, if you are set to work on uncongenial tasks, poorly suited to your personal skill-set and personality, then no matter how positive the surroundings, you will begin to fret and stop giving of your best. You need to be set tasks at which you believe you can excel at, and work in a positively charged atmosphere that helps you to succeed. Your own satisfaction and pleasure will help motivate others and generally reinforce a positive feedback loop to the general benefit of the organisation.

The essential point is that you are all individuals with different personalities and personal needs that overlay Maslow's pyramid. You can see this even at the basic needs level: imagine two very similar people with similar skills doing similar work.

One might have a non-working partner and three or four school-age children; the other might be a bachelor with a penchant for mountaineering. The father will need a family home; the other a small apartment. The father will need a high income to pay the myriad of bills; the other might be happier with a smaller salary but more time in the mountains. It follows that, as a team leader, you can exploit even the most basic needs to increase your reports' motivation, in this case by devising creative 'cafeteria' rewards schemes, which allow individuals to make up a total package which most closely satisfies their personal needs and aspirations.

More generally, personalities differ: at one extreme are hermits who are perfectly happy gaining satisfaction from their work with minimal human intervention; at the other are people who are happy only when interacting with other people. Most of us are somewhere in between, able to spend periods concentrating on work but needing social interaction. It is part of the team leader's job to analyse the job/people make-up of tasks and then to assign them to individuals whose personalities most closely approach the required mixture.

Much unhappiness is caused by failure to observe this simple rule, usually because you are so concerned about matching the technical knowledge or practical skill requirements that you forget that you are, for example, asking a 'hermit' to front customer services, a job for which, no matter how excellent their product knowledge, they are temperamentally quite unsuitable. (The impact on the individual is to provoke fear, and hence undermine their need for security, thereby breaking the 'Maslow rule' that high-priority needs must be satisfied first.)

Another important aspect of this theme of personal needs as satisfied through the workplace is that people grow up and, as they mature, their interests evolve. This is part of the iterative process of seeking new challenges when old challenges have been exhausted. As a team leader you must recognise when people are ready to move on, and help them prepare for the migration by introducing them to appropriate training and opportunities in areas for which they are now temperamentally and intellectually ready.

For example, many young people find computers fascinating in themselves and gain rich fulfilment in programming. For some, programming seems never to grow dull, since there is always a new problem to solve or better techniques to be designed, but many people 'grow out of' programming. Where do they go? There is no one simple solution again, since they are all different individuals. Options might include the following (the directions do not imply value judgements):

- forwards towards systems and applications design
- back towards manufacture, engineering and design of the hardware
- directly towards sales or marketing of the machines or the software
- sideways towards training and customer support
- upwards towards management of programmers

In the course of a career, either as a result of natural development or as a result of an inspired manager asking you to use existing skills to tackle a 'new' problem or aspect of the business, you can pass through many different phases, adding new skills and developing a higher-level understanding of the products and their capabilities, hence adding more value to the employer's business.

Another important personal need is increased responsibility, since the granting of responsibility is evidence of esteem. You know this from personal experience: when you begin to work at a new type of task your teacher or manager hangs over your shoulder to correct your errors and guide you round pitfalls. Soon you feel confident enough to work without such close supervision, and begin to resent it, feeling that 'they don't trust me' and that you are held in low esteem. If, on the other hand, supervision is tapered off or withdrawn just before the manager is fully confident that the new lessons have been fully assimilated, then the message is 'Wow! They trust me, even though I am not really quite ready', a powerful stimulant to your motivation.

Traditionally, in highly stratified organisations with many management layers, adding responsibility was easily done by an annual or biennial promotion. Nowadays more subtle approaches are needed, such as:

- withdrawal of supervision
- 'bigger' jobs of the same type
- wider responsibility within the same job (e.g. adding stock ordering to an assembly team or making a customer service team partly responsible for defining product enhancements)
- adding training and supervision of new recruits to the 'doing' part of the job

These methods of injecting additional responsibility into teams doing basically repetitive work are familiar to students of Japanese management techniques; there are many motives for following this path, but demonstrating trust in and raising the self-esteem and self-fulfilment (and hence the motivation) of the work group is an essential element.

7.4   Management and motivation

Apart from the two management techniques described above, properly matching jobs to people and increasing delegated responsibility, the team leader is responsible for the motivation of their teams in two ways:

- by providing a working environment conducive to high morale
- by providing leadership

Academic studies have come up with an interesting fact about working environments. They rarely help increase motivation, but they can and do decrease it. Many of us have at some time worked in superficially poor environments: overcrowded, poorly equipped or otherwise uncongenial (and even if we have not, highly motivated miners or soldiers certainly do!), but if the human atmosphere is good the physical aspects become irrelevant. However, if the atmosphere falters, then the environment often takes the blame. A frequent cause for 'environmental' dissatisfaction is petty bureaucracy, the imposition or enforcement of time recording and similar rules and regulations which suggest that we are not trusted, and all of which lower our self-esteem. So a team leader can improve the working environment, from a motivational viewpoint, by continually asking staff what they most dislike and how they would like it changed, consider the implications and where possible implement the changes suggested.

If you think for a few moments about your own working environment, physical and emotional, you may make an interesting discovery: many of the changes would cost little if anything in money terms to fix. Amongst the most annoying elements are things that don't work properly yet which are not brought back to full working order. Obvious examples are flickering fluorescent tubes, doors or drawers which will not open or close easily and chair castors which stop running 'true'; yet how often does a handyman appear, charged with spending an hour fixing things? A regular (timetabled, so that absentees can leave requests) visit from a handyman must surely be the most motivating way of spending £50 yet devised, taking just a few minutes to remove sources of irritation on the one hand and giving a practical demonstration that you care about your team on the other. (It will also extend the life of company assets and help prevent that build-up of embarrassing broken bits and pieces stacked up behind a screen in a distant corner.)

As an aside you might note that improvements in the physical environment can easily be counter-productive in motivational terms. A common example is redecoration. All workplaces eventually get grubby, carpets wear, paintwork gets chipped and stained, furniture gets damaged, and so on. The organisation quite rightly decides to give the place a face-lift. But who decides the new décor: the boss (or his wife)? An expensive bought-in specialist, fully up to date with interior design trends and the latest fads of occupational psychologists? Or the people who have to work there? Do you remember a workplace makeover when the immediate and universal reaction was 'Yuk! Who on earth thought up this nonsense?' How much better to show your team members you trust them and let them make the decisions. They do, after all, have to live with the results, and are quite likely to choose cheaper and more practical options than 'professionals'. Similarly with workplace layout. It might look nice and tidy to have regimented rows of desks but, left to themselves, they are likely to come up with something much more practical and convenient for everyday operations, as well as suiting their individual personalities and needs for privacy or company.

The second managerial input is 'leadership'. The primary job of the team leader is to ensure the long-term wellbeing of the organisation through the application of available resources to corporate and departmental objectives. In simpler words: using people to get there. A well motivated workforce with clearly defined and properly understood objectives will get there on their own without much management intervention. Various aspects of leadership are considered in other parts of this book, so here is a brief summary:

- Enthusiasm for the goals and the means by which they are to be achieved. Enthusiasm is contagious and one of the most important motivating tools in the team leader's armoury (equally, conspicuous lack of enthusiasm is the quickest way to stop a well-motivated group dead in its tracks).

> 'Feedback is the breakfast of champions.'
>
> *Ken Blanchard*

- Courage, and particularly the courage always to be honest and open with bad news. You may have to fight on behalf of your people for resources or cooperation from other groups, and to let them know you have tried and succeeded or tried and failed. Build a climate where you are trusted.
- Praise for good work, offered sincerely, spontaneously and frequently raises the self-esteem of team members, makes them feel good about themselves and hence motivates them to do better as well as to keep you informed of progress and potential problems.
- Trust your team and delegate responsibly. You know they can do it so let them get on with it, taking the necessary decisions along the way; back up delegation with an immediate non-judgemental response to calls for help; treat their mistakes and shortcomings as positive learning experiences rather than excuses to issue reprimands; treat people as individuals with personalities as well as skill sets; involve people in setting objectives and measurement criteria, assessing their future growth and consequent training needs. (See Chapter 8 for more on appraisals.)
- Open and frequent two-way communications; if a choice has to be made, ask those affected for opinions and advice, and accept and act upon that advice unless there is an overriding reason to do otherwise, in which case give a full explanation for your decision. Knowledge may be power, so empower your people by sharing information as soon as it becomes available (keeping 'secret' information readily available on the workplace 'grapevine' is an excellent way of eroding trust and undermining delegation).

Some of these characteristics overlap or are restatements of common themes, but they are observed in most true leaders. A few people have a natural instinct for raising morale and creating positively motivated teams, no matter how adverse the circumstances. The rest of us will find that, if we actively go about reducing 'negatives' in our own attitudes and behaviour and follow the guidelines given in this section, we will soon be seen to be 'above-average' team leaders, even if we cannot claim to be 'inspirational' – and even that may come with practice!

7.5 'The proof of the pudding!'

The old saying, 'the proof of the pudding is in the eating' applies as much in the area of staff motivation as in the kitchen. There can be few people who doubt that high levels of motivation make an important contribution to direct productivity as well as providing indirect benefits, such as staff retention and speed of reaction in a rapidly changing world. You have seen some things which motivate us all as individuals and some techniques you may use to raise motivation in others. Now all that remains is to use them.

Many companies, as well as organisations from football clubs to the armed forces, have grasped the opportunities and faced up to the challenges of introducing what in some cases has been a complete change of internal culture. Those who succeed in maintaining high levels of motivation over the long term remain successful through the peaks and troughs of the economic cycle, no matter the pain inflicted by recessions and other adverse circumstances. The spirit of the workforce enables them to pull through, while poorly motivated organisations seem to shrivel up and die.

3M Corporation, Motorola and the Sony Corporation are just three companies who encourage high morale by encouraging employees to take risks, to have ideas and bring them to fruition through their 'entrepreneur' programmes. No blame attaches to failure, provided all concerned learn from the experience, and success is rewarded; the organisation and the individuals each gain from all manner of experiences.

The primary rewards are the esteem of colleagues and superiors, self-fulfilment, additional responsibility and, at the bottom of the list, sometimes financial (or non-financial) benefits, depending on the personal needs of the individual. This is important since, as we have seen several times in this chapter, the actions needed to raise motivation are often cost free or, at most, low cost. The returns to the organisation can be high, but the 'cost' to management is just kicking aside mental barriers erected in the past, and cutting loose from the chains of habit and fear of change.

7.6   Chapter summary

- Remember Maslow's Pyramid of Needs and how basic needs have to be satisfied before higher-level needs are met.
- Recognise and reward members of your team according to their individual needs and requirements.
- Fit the personality to a job, and recognise when team members are ready to expand and develop their workload, offering appropriate training if required.
- Recognise and acknowledge a person's need for self-esteem, and show you trust them.
- Your style of leadership can really make a difference to the performance of the team.

Chapter 8

Giving and receiving feedback and the appraisal process

This chapter will examine the setting and monitoring of performance standards, clarifying the need to balance task and people management in your role as a team leader. It will move on to explore giving and receiving feedback as a key factor in managing performance, and how to deliver a formal appraisal process in a structured and productive way.

8.1   Setting and monitoring performance standards

Looking at how you should manage the performance of team members it is important to consider the bigger picture. You need to understand how performance standards should bear a straight-line relationship to what your organisation has set out to achieve. The diagram below clarifies this.

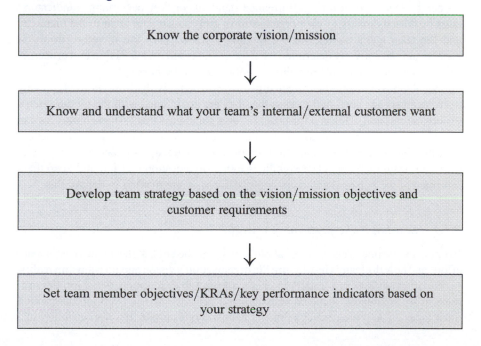

Know the corporate vision/mission

↓

Know and understand what your team's internal/external customers want

↓

Develop team strategy based on the vision/mission objectives and customer requirements

↓

Set team member objectives/KRAs/key performance indicators based on your strategy

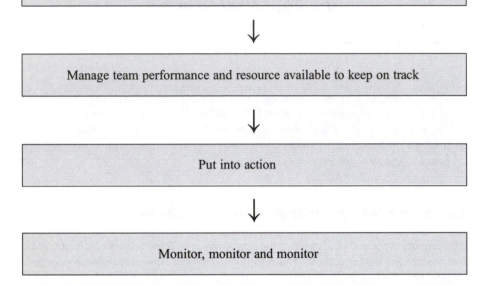

As can be seen, your team's performance standards or 'key performance indicators' must be related directly to your organisation's aims above your team in the hierarchy and the team members responsible for them, below. They are therefore firmly *in context* and can correctly determine what training, resources, leadership and processes are required and even what recruitment policy should be pursued.

The obvious connection between organisation and individuals is crucial. Too many team members are 'lost' in the sense that they are not aligned with the organisation's vision and mission. This is a mistake. Today's paradox is that just as employees can expect less loyalty from organisations, these organisations are more than ever dependent on high-level performance from employees. One way to generate loyalty is to give your team members clear sightlines to the organisation's vision and keep them informed of the personal and team contribution they are making to deliver this.

8.1.1   So what is the purpose of my job?

Every team member must be clear about how he or she contributes to the overall team effort, and how the team's results turn the organisation's vision and mission into reality.

We often advise team leaders to create a team job description. Most job descriptions, if we are honest, focus on activities and behaviours. The team job description must focus on activities, behaviours *and* results. For example, if a responsibility in a job description is to 'produce monthly reports' the question to ask is 'in order to do what?' The point is to keep asking the question 'in order to do what?' until you finally reach a concrete result which contributes towards overall organisation success.

8.1.2   Key results areas (KRAs)

KRAs flow from individual or team job descriptions. KRAs contribute towards clarifying 'how should we (as a team) be spending our time?' For the team's strategy to succeed, all team members must achieve specific results which, when accumulated, ensure that all strategy elements are delivered. KRAs are a summary of those key factors of individual or the team's job that are vital for the team's success. KRAs must be monitored monthly as a minimum. Otherwise, divergences from your strategy may occur.

If you have difficulty identifying your KRAs ask yourself the following question: what are the major aspects of the team's work that could go wrong? You might say 'we don't realise our sales target', 'we use up our resources', 'too many customers complain' or 'we don't complete our reports on time'. These can be translated into the following KRAs – sales, finance, customer service and monthly reports.

8.2   Reviewing and evaluating performance

People in management as a group tend to be uncomfortable with the human aspect of performance, and prefer to concentrate on more familiar technical matters. However, an essential part of your role as a team leader is to be able to communicate effectively, motivate, boost morale and develop others. Reviewing and evaluating the performance of staff and helping each individual to expand skill sets and competencies will ultimately improve organisational performance and feed into business planning. Giving feedback should be part of your everyday activities, with the key milestone of a formal appraisal, usually undertaken on an annual basis, to review a team member's performance, establish specific goals and objectives and discuss career interests, development needs and requirements.

Why bother? Because one-man teams don't work. Your job is to coordinate your team's effort to achieve organisational objectives. This is a tough job. Increasing productivity and adding value/profitability to organisations constantly having to reinvent themselves due to the pace of change doesn't come easy. People get stressed, distracted, confused. Some jockey for position; some give up. They look at *you* to 'fix things', and even those who are willing to help can't agree on *how* things should be fixed. This is your challenge.

8.3   Creating balance

As a group, those in your team need to understand not only the task you have allocated to them, but also how it fits into the bigger picture of activities going on in the team.

Life as a team leader is a constant balancing act (Figure 8.1), getting the right emphasis between task and people management. An imbalance between the two

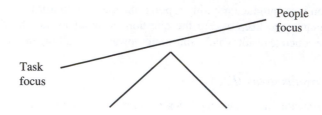

Figure 8.1   Creating balance

will affect team spirit and cooperation and could, for example, spoil the chances for success during a project.

Too much *task* focus can show up as:

- pushing for results too soon
- too much concern for efficiency and time
- one or a few members planning experiments outside of team meetings
- assignments handed out to other team members outside of meetings

This could result in:

- concerns being ignored
- limited participation of some members
- discussion in meetings dominated by a few people
- decisions that are made based on opinion, rather than based on facts provided by subject matter/technical support

Too much *people* focus can show up as:

- too much concern for feelings and personal problems
- letting discussions wander until everyone airs all their fears
- trying to do everything by consensus decision-making

This could result in:

- project progress suffering
- no forward progress
- no milestones met
- action items repeated or continually extended

8.4   Giving effective feedback

The purpose of feedback is either to maintain or to change performance, in order to keep an individual or team on track to achieve their work goals. Feedback should be viewed as a way of giving help. The trouble, however, is that it is often perceived as a threat or an attack by the recipient. So you need to be sensitive, informative, and aware of the pitfalls that can occur during the process.

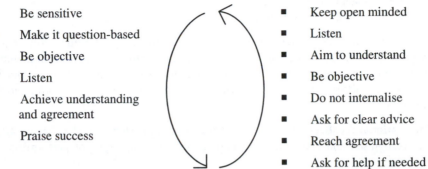

Giving		Receiving
■ Be sensitive		■ Keep open minded
■ Make it question-based		■ Listen
■ Be objective		■ Aim to understand
■ Listen		■ Be objective
■ Achieve understanding and agreement		■ Do not internalise
		■ Ask for clear advice
■ Praise success		■ Reach agreement
		■ Ask for help if needed

Figure 8.2   The two-way approach to giving effective feedback

Let's start with some basics. The two-way process should be as shown in Figure 8.2. This process should never be entirely one way, from the giver (sometimes felt to be 'superior') to the receiver (sometimes felt to be 'subordinate'); it should be an exchange of objective views, facts and guidance between two adults conferring on a point of mutual help for the good of both people.

Our definition of feedback is that it is communication to a person (or group) which gives that person information about how s/he affects others. As a guided missile system, feedback helps keep an individual 'on target' to achieve agreed objectives.

How you pass on information or communicate information will depend on the nature of work that you do. It's easier for a team leader or manager to give (negative) feedback than a member of the team. If you report into a manager, and want to give (negative) feedback choose your time and place carefully! A neutral venue – not *their* office – and during an appraisal session or one-to-one meeting is usually best.

The giving (and receiving) of feedback requires courage, skill, understanding and respect for self and others.

When giving feedback there is ONE rule:

> When giving feedback, DO
> consider the person's capacity to ACCEPT your feedback

When receiving feedback there is ONE rule:

> When receiving feedback, DO
> not take it personally!
> It is another person's view at one point in time and will change based on what
> you do with the information. Ask yourself: are they telling me this to
> help or to hurt me?

- Knowledge of 'self' is dependent on you (the person giving feedback)
- Use 'I' statements rather than 'you' or 'we'
- Never be evaluative or judgmental

When you delegate a task, for example, asking a member of your team to write a report or assess the viability of developing a new piece of equipment or software programme, there needs to be a regular review process in place to ensure that the person is 'on track'.

There are different types of feedback which should then be used according to the situation:

Motivational feedback tells a person that their good performance has been noticed and recognised and gives them the impetus to repeat this type of performance in the future.

Feedback should be given as soon as possible after the event/activity has taken place. Be careful in this situation of giving extra advice or suggestions for even better performance at the same time, as this could be interpreted as criticism.

Development feedback indicates to a person what needs to be improved and asks them, for example, how they believe they could have tackled a task/situation in a different way in order to learn for the future. If you have to correct someone's perform-ance do not do it in front of the rest of the team; whenever possible do it in private.

Correcting small issues like the layout of a proposal or modification of a task is usually quite straightforward. Needing to review a person's behaviour can be more challenging, so let's look at this in more detail.

8.4.1   Giving and receiving feedback on behaviour

Some of the most important information we can receive from others (or give to others) consists of feedback about how we behave. Such feedback can provide learning opportunities for each of us if we can use the reactions of others as a mirror for observ-ing the consequences of our behaviour. Such personal feedback helps to make us more aware of *what* we do and *how* we do it, thus increasing our ability to modify and change our behaviour and to become more effective in our interactions with others.

You should focus on:

- **Behaviour rather than the person**
  It is important that you refer to what a person *does* rather than comment on what we imagine s/he *is*. This focus on behaviour further implies that you use adverbs (which relate to actions) rather than adjectives (which relate to qualities) when referring to a person. So you might say a person 'talked extensively in this meeting', rather than this person 'is a loudmouth'. When you talk in terms of personality traits it implies inherited qualities that are difficult, if not impossible, to change. Focusing on behaviour implies that it is something related to a specific situation that might be changed. It is less threatening to a person to hear comments about their behaviour than their 'traits'.
- **Observations rather than inferences**
  Observations refer to what you can see or hear in the behaviour of another person, while inferences refer to interpretations and conclusions which you make from

what you see or hear. Remember these inferences or conclusions may be wrong. Where they are shared (and it may be valuable to have this data), it is important that they are recognised as inferences, not facts.

- **Description rather than judgement**

  The effort to describe represents a process for reporting what occurred, while judgement refers to an evaluation in terms of good or bad, right or wrong, nice or not nice. The judgements arise out of a subjective, personal frame of reference or values, whereas description represent *neutral* (as far as possible) reporting. Avoiding evaluative language reduces the need for the recipient to react defensively.

  It is difficult not to judge or evaluate people. We grow up in a world where most people do it, and it tends to give us a sense of superiority or rightness to evaluate what others do. One thing which may help to counteract this is to separate what the other person is doing from what you feel as a consequence. So, 'you're a bore' probably means 'you talk a lot and so I feel bored'. Putting this the latter way is much more helpful since the receiver of the feedback may learn that he needs to talk less, but the giver may learn that he easily loses interest when other people are talking.

  This may be helpful when receiving feedback. When being judged critically, people either accept it and blame themselves, or reject it and say others don't understand. However, if you can separate the judgement into 'what am I doing?' and 'what is s/he feeling?' the feedback becomes easier to understand and may lead to both parties learning from the event.

- **Behaviour related to a specific situation, preferably to the 'here and now' rather than to behaviour in the abstract, placing it in the 'there and then'**

  What you and I do is always tied in some way to time and place, and we increase our understanding of behaviour by keeping it tied to time and place. Feedback is generally more meaningful if given as soon as appropriate after the observation or reactions occur, thus keeping it concrete and relatively free of distortions that come with the passing of time.

Dimensions of feedback

Here and now	**There and then**
'My experience of you now is . . .'	'At that time I felt you were . . .'
Implicit	**Explicit**
Often non-verbal	Tone is clear and direct voice, process level.
Evaluative	**Descriptive**
'You kept interrupting and you kept scratching your head.'	'That was bad.'
Quantitative	**Qualitative**
'You interrupted 20 times.'	'You interrupted a lot.'
Group	**Individual**
'The group lost direction.'	'You, Bill, lost direction.'
Positive	**Negative**
'I liked it when you . . .'	'I didn't like it when you . . .'

- **Sharing of ideas and information rather than giving advice**
  By sharing ideas and information you leave the other person free to decide for themselves, in the light of their own goals in a particular situation at a particular time, how to use the ideas and the information. When you give advice you tell the others what to do with the information, and by doing this you take away their freedom to determine for themselves what is for them the most appropriate course of action.

- **Exploration of alternatives rather than answers or solutions**
  The more you can focus on a variety of procedures and means for the attainment of a particular goal, the less likely you are to accept prematurely a particular answer or solution which may or may not fit a particular problem. Many of us get around with a collation of answers and solutions for which there are no problems.

- **The value it may have to the receiver, not the value or release that it provides the person giving the feedback**
  The feedback provided should serve the needs of the receiver rather than the needs of the giver. Help and feedback need to be given and heard as an *offer*, not an *imposition*.

- **Behaviour that the receiver can do something about**
  Frustration is only increased when a person is reminded of some shortcoming over which s/he has no control.

- **The amount of information that the person receiving it can use, rather than the amount that you have which you might like to give**
  To overload a person with feedback is to reduce the possibility that s/he may use what s/he receives effectively. When you give more than can be used you may be satisfying some need for yourself rather than helping the other person.

- **Time and place so that personal data can be shared at appropriate times**
  Because the reception and use of personal feedback involves many possible emotional reactions, it is important to be sensitive as to when it is appropriate to provide the feedback. Excellent feedback presented at an inappropriate time may do more harm than good.

- **What is said rather than why it is said**
  The aspects of feedback which relate to *what, how, when, where* of what is said are observable characteristics. The *why* of what is said takes us from the observable to the inferred, and brings up questions of 'motive' or 'intent'. Making assumptions about the motives of the person giving feedback may prevent you from hearing or cause you to distort what is said. In short, if I question 'why' a person gives me feedback, I may not hear *what* he says.

Follow this by:

- **Checking the communication has been clear**
  One way of doing this is to ask the receiver to try and re-phrase the feedback s/he has received in his/her own words to see if it corresponds with what you, the giver, had in mind.

- **Checking the accuracy of the feedback**
  When feedback is given in a work group, it gives all parties time to assess how accurate they believe the feedback to be. They can come to the conclusion as to whether it is one person's impression or an impression shared by others.

Feedback, as stated earlier, is a way of giving help; it is a useful mechanism for the person who wants to learn how well their behaviour matches their intentions; and it is a way of helping an individual or group establish their identity.

8.4.2   Ways in which people can respond to feedback

They can:

- accept
- exaggerate
- reject the feedback and/or the giver of the feedback
- seek clarification
- allow in and consider
- get confused
- feel hurt
- be surprised
- feel guilty
- feel satisfied
- feel delighted
- become withdrawn
- look for support

8.4.3   How to set up feedback

1. Time out, i.e. stop the activity at a mid-way point to allow time for review either in full or as a small group.
2. Individual review, i.e. concentrating on one group member.
3. Establish guidelines with regard to positive and negative feedback.
4. Allow feedback based on a clear distinction between observation and intuition/ interpretation.
5. Get observers to establish observation criteria, i.e. decide what they are going to look for.
6. Get the task group, individually or collectively, to establish the criteria for the assessment of their performance.
7. Make the content of the exercise the basis for the process review (e.g. agree and put in priority order five criteria for assessing group effectiveness).
8. Use closed-circuit television and/or tape recorders. Perhaps allow a task group to review/analyse and present to the full group.

Guidelines for giving feedback	
Giver	**Receiver**
Address the person directly.	Treat the feedback you are receiving as information.
Be specific about what aspect of the individual you are giving feedback on.	Make your own choice about what you intend to do with the feedback. (Accept, reject, wholly or partly.)
Be clear about your own reaction to the other person.	Let the other person know what you are doing/have done with the feedback.
Offer your feedback as information without attached conditions.	Avoid arguing, denying, justifying, swallowing whole.
Offer as much as you think will be useful. Avoid giving a long list.	
Make a clear distinction between observation and evaluation.	Ask for clarification if you do not fully understand the feedback.
Remember to give positive as well as negative feedback. Be sensitive to how the other person is reacting to your feedback.	Distinguish between the content of the feedback and your reaction to it.

8.4.4   *Questions/comments which can be used in review session*

- Is there anything you wish you had handled differently?
- What were the consequences for you of. . .?
- What did you expect to happen?
- Have you had feedback like this before?
- What were you doing?
- What did you see others doing?
- Can you be more specific about. . .?
- How did you stop yourself from. . .?
- What, as far as you're concerned, was the worst thing that could have happened in the exercise?
- What are you feeling now?
- What did you achieve?
- What are you doing with the feedback you have received?
- What did you like about what you did?

Feedback is obviously a key communication activity in performance management. Moving on from this, there needs to be a more formalised review process in place to assess each individual's performance against objectives and standards for the trading year – the annual appraisal.

Very often this form of performance review is seen as an administrative chore and is dreaded by both parties concerned. One of the ways to make the appraisal process more palatable is by meeting each team member individually and regularly through-out the year, say on a monthly basis, for one-to-one small review sessions. Meaningful regular discussions about work, career, aims, sense of direction, dreams, life, even last night's TV, help build trust and understanding between both parties, and reduce the possibility of needing to raise a corrective performance issue during a formal review session. This will have been 'nipped in the bud' earlier, and allows the annual appraisal to be more about creating a productive future.

8.5   The appraisal process

Overall, the intention of the appraisal process is to act as a catalyst to make future performance better than, or simply different from, that of the past. This may some-times involve identifying and correcting personal weaknesses, but – just as likely – environmental changes or an organisation's intentions make changes to what and how things are done necessary. Appraisal must not be a witch-hunt; the focus should always be positive and on the future.

Every organisation has a different way of conducting performance appraisals and you will need to check with your HR department as to which method is used in your company. There could be a web-based template that you download, or you just type up a report for your manager on an annual basis. One way or the other you need clear guidelines as to what is expected and what should be produced.

So, consider the purpose. Appraisal should act to:

- review an individual's past performance
- be transparent, putting in place performance measures for the short, medium and long term
- clarify, define and redefine priorities and objectives
- plan future work and the role development
- agree specific goals
- identify personal strengths – including unused hidden strengths
- identify areas for improvement
- create a 'whole person' training plan
- build a mutual understanding between you as the team leader and the team member
- act as a catalyst to delegation
- reinforce and cascade the organisation's philosophies, values, aims, strategies and priorities

These intentions are not mutually exclusive, though the emphasis in an individual appraisal may be more on some than on others. Your role during appraisal is actively to contribute to your organisation's objectives on a broad front, using your report's skills for the purpose. The team member's role is to use their skills to help management achieve the organisation's aims. The appraisal session is a time to synchronise the two.

8.5.1   Clear objectives

Every appraisal scheme must have clear intentions. To achieve this and direct discussion, many organisations have a prescribed approach: this usually lists areas of planned discussion, and includes an element of evaluation to measure past performance objectively.

Good documentation can act as a universal basis for all appraisals, but details appropriate to individual jobs may need accommodating with some amendment. The intention with any guidelines is to prompt an effective and systematic approach, not to be a straightjacket and hinder a flexible approach. This documentation should have been supplied to the appraisee well in advance of the appraisal meeting.

Any 'score' in the system acts as a means to an end, existing primarily to prompt action. Take a common necessary skill such as report writing. If someone is less than skilled at this (and marked 'below average'), what really matters is that action results and improvement follows. The organisation can then reap the advantages of their enhanced ability. Scores are only a prompt to create specific action – ranging from individual counselling to attending specific skills development training.

8.5.2   Preparation

In readiness for an appraisal, both parties should have reviewed statements of objectives agreed at their previous meeting and should prepare positive discussion points arising from the extent to which these objectives have been met or exceeded. (We can assume that if a significant shortfall is expected, warning bells would have been sounded a long time ago, and remedial action taken or the objectives renegotiated.)

You should have reviewed the previous appraisal form documentation (if appropriate), to identify possible trends in performance or behaviour and judge if they still apply. You should think through the areas of performance that can be singled out for praise, the questions to be asked, plus any information that needs to be brought up at the meeting. If there is to be an organisation-initiated change/divergence you might like to have some ready-prepared evolutionary options to feed into the later stages of the discussion.

Preparation for the person being appraised means thinking over the past period, to review in their mind if and how they have achieved objectives highlighted during the previous appraisal. If there is any shortfall in performance they should decide how they might remedy the situation. In turn, they should have been encouraged by you to consider career developments that they might wish to make.

A vital element of preparation for both parties is to take time to sit back and contemplate the appraisee's contribution over the entire period considered during the review portion of the meeting. It is most important that problems, set-backs or upsets during recent weeks should not cloud perceptions of year-long work.

8.5.3   Setting up and running the right kind of meeting

The meeting itself needs some care. You as the appraiser should bear in mind that:

- The meeting and the agenda should be set up well in advance (and any necessary documentation read ahead of the meeting).
- Adequate time should be allowed – often 90 minutes to three hours (results potentially making this time well spent).
- Surroundings must be comfortable and interruptions *must* be prevented.
- All those involved must agree about the format and the practicality of the proceedings.
- Documentation and any element of 'scoring' must be made clear in advance.
- Targets and other objectives relevant to the period under review must be to hand; every aspect of the appraisal must deal with facts (not judgements made on hearsay or uncertain memory).
- Discussion must be open, judgements objective and everyone open minded.
- All should recognise that listening is as important as talking.
- Sensitive issues must be tackled; peer embarrassment must not sideline them (criticism is part of appraisal, though it must be constructively given and received and lead to change if necessary).

Additionally, two overriding principles are paramount. In a successful appraisal:

- The appraisee does most of the talking (though you as the appraiser chair the meeting).
- The focus and weight of time and discussion is on the future more than the past (the two go together, of course, but the end results are action for the future).

A clear agenda is vital. Stick to it, and avoid digressions. If new topics or opportunities for creative discussion appear, schedule a further meeting and do not upset the appraisee by letting time pressures squeeze listed topics they regard as important off the agenda.

A positive atmosphere during the meeting

Let's make a couple of reasonable assumptions:

- The appraisee has been working in the context of an explicit job description covering functions and responsibilities.
- The session is the culmination of a period for which concrete appraisee objectives were set and mutually agreed at or since the last review.

So begin by handing out some praise and creating a positive atmosphere.

Note that praise is given for what the person has actually done, whether or not targets were met. So, for example, if a sales engineer has not met sales targets in volume or value terms, it may be possible to praise exemplary sales presentations and written proposals. Subsequent discussions may then focus on helping your team member to improve and monitor his or her progress. Your goal is to work to your team member's strengths so that they can be leveraged to assist your team overall.

If the appraisee hasn't done anything particularly well this will not be possible, and you will need to take stock of the situation and recommend appropriate action. It may be that this person is new to the role and it has simply turned out that they are a round peg in a square hole, and you should review other roles and activities that might be more appropriate for them. Or maybe an able and well-respected long-term member of the team has been transferred into an unsuitable situation, and should be redeployed. Or perhaps family problems have interfered with work and consideration should be given to leave of absence or counselling.

Obviously individual cases require individual and sensitive handling in an atmosphere of trust and complete confidentiality, and both of you must recognise that an appraisal session is not appropriate for this level of discussion.

So working on the premise that the person has performed well, kick off the session with sincere praise for notable achievements and then get the appraisee to review their performance by asking questions such as:

- What do you think were your best achievements?
- What gave you particular satisfaction?
- Of all the things you have done recently, which gave you greatest pleasure?
- What are your strong points?

These questions give both parties an agreed picture of how the appraisee has evolved since the last session.

They may have tried doing a few new things, such as addressing a conference, giving a training course, or helping to plan or lead a project rather than just executing plans devised by others. If these rated high on the satisfaction/pleasure scale, we have measures of recent personal development and clues to future evolution.

The same applies if some long-term activity has metamorphosed from being a chore into being a source of delight.

Remember the past!

We all change with time, and during this part of the appraisal the team leader should have easy access to the notes made during previous sessions, so as to spot trends. For example, many of us were very nervous the first time we addressed a large audience and probably hated the experience, but were forced by circumstances to try again and found it easier and easier, and suddenly discovered that it was fun and a source of pleasure. A succession of meeting notes should make this clear and the organisation has now gained a willing representative for the conference platform or classroom. Evolution at work!

Afraid of change?

So now is the time to turn from the past and present towards the future. In general, since we tend to fear change, most of us will seek to continue doing our present jobs with some fine tuning of the content or training so that we can do these better or specialise in the areas we do best. However, keep the discussion content flowing from the appraisee:

- What tasks can you handle in the future?
- In which other areas can you improve your performance?
- What objectives would you like to set for yourself?
- What are the timescales for these?

Assuming that both you and the member of your team you are appraising expect the same relationship to continue, albeit with some evolution in the actual work content for the next period, you can move on to agree some concrete, measurable, win–win objectives. You should takes the lead by asking questions such as, 'How shall we measure your performance against your targets?' and 'What will you achieve by X date?'

In this scenario, you will be mentally matching the appraisee's desires against departmental and corporate needs in order to ensure that the individual is able to make a positive contribution towards pre-established (even though changing) goals. It is also necessary to caution then against over-stretching themselves or underestimating their abilities, since both extremes are damaging to the individual and could be harmful to the organisation. Eventually both parties should agree on sensible SMART (Specific, Measurable, Agreed or Achievable, Realistic and Timed) objectives, write these down and agree them. Then put them in a safe place for a day or two.

We can all get carried away by the euphoria generated by a happy and productive meeting, so it makes sense to allow these couple of days for personal review of what has been agreed. Is the target really too demanding? Arrange a few minutes to review the agreement in the cold light of a new day and then sign these or adjust before signature. Have a copy each, along with copies of any forms or minutes of the meeting, so that they are always available for reference and can, if need be, act as an objective basis on which to discuss divergences between promise and outcome in the future.

Divergences (not negatives!)

Throughout the last section there has been an underlying assumption that the period following the appraisal would be pretty similar to the period preceding it. The department's and team's objectives would remain much the same, the same team leader– team member relationship would continue, and the person would carry on making the same contribution. Very often this is the case. However, there are a variety of factors that can change this assumption and if we are to be realistic we should consider some other cases.

Let's say this person is valued for their present and potential contributions and we wish to keep them in our team. Since the appraisal is specifically designed to be centred around the appraisee, what negatives or divergences might emerge?

Here are some examples of what they might say:

- My personal circumstances have changed and I need to spend more time with my family; as much as I enjoy business travel I want less of it (or the opposite).
- Thank you for saying I am a good engineer/instructor, but I am getting bored with the xxxxx aspect of the task; I want to move towards x/y/z.
- It has been great working in the field/headquarters, but now I want to gain experience in headquarters/the field.
- I have had a great time working here, but I (or my partner) feel that this would be a good time for an assignment in another country (or to go home).
- I consider that I am good at my current job, but think I am ready to move to a more senior position.

The appraisee is not unhappy, but feels that the time has come to plan for change, either to evolve their career, or to take account of events or pressures from outside the workplace.

Now we have come so far in a very positive atmosphere, so we must continue this way even though, as a team leader, you may feel disappointed or hurt by what may seem like rejection from the person you are appraising.

Two observations may help you deal with this:
1. We all went through the same developmental process as we outgrew a succession of jobs in our own pasts. We didn't in some way reject or even criticise a previous manager who had helped us grow until we reached the stage of wishing to move on. Personal change and growth is natural.
2. As a team leader you have two roles: you run your team, but you are also a representative of the organisation as a whole; when conducting an appraisal you are in part seeking to promote the health and wealth of the entire organisation, even at a short-term cost to your own team.

Now it is obvious that in most cases you will not be able in the short term to do more than accept the inputs and promise (sincerely) to do whatever you can to accommodate the wishes of this person. That is sufficient. You then have to carry out your promise and, if necessary in consultation with the HR Department and managers running other departments, plot out some possibilities for discussion with the appraisee at a continuation of the session.

Changes in the organisation may well be emerging, and you may have been told to help implement these changes. For example:

- The company has set up a local sales or support organisation in country X, so we don't need to keep flying you to and fro (or we want to set up an organisation in X and need people on temporary assignment to get it going).
- The ranges of services and support facilities are evolving and we need to build up expertise on the new and de-emphasise the old.

- We are refocusing our approach to the market and need people to help manage the transition.
- Our team has established a best practice approach in a particular area of activity, and we have been asked to raise awareness in other units.
- Some other department needs our expertise to help them.

As a result of this corporate evolution there may now be opportunities that you should bring to the attention of the appraisee, showing how their expertise could bring greater value to the organisation by working on a different stage.

On the other hand, the news may not be positive, such as the department or your team is to be run down or their job is to be phased out. If this is the case, you need to have explored all possible options for redeploying this person with your manager as well as consulting with your HR department well in advance of the appraisal meeting, so that you can discuss the appropriate way forward during the review.

8.5.4 After the meeting

Some difficulties may occur, but the overall feeling and outcome should always be constructive. If individual weaknesses are identified they will need to be addressed. If changes demanding personal development occur, then priorities must be set and action implemented. If new roles or responsibilities are agreed, they will need formalising. Experience shows that only a poor or unusual appraisal does not produce useful action for the future – and how many new ideas are needed to justify the time taken?

Despite other pressures, the worst possible outcome is for an appraisal to set something useful in train but not follow it through, so always record and follow through decisions. Even the best performances can be improved. Even the most expert and competent people have new things to learn and new ways to adopt; the present dynamic work environment sees to that. The appraisal process is *not* all about criticism and highlighting errors or faults (though realistically there may be some of this). It is about using analysis and discussion to move forward. It is about building on success, sharing good experience and effective approaches and, above all, it is about making more of the future.

And if you, and your people, are not there to affect the future, what are you there for?

Case study: Demanding performance standards[1]

Certain organisations have a very rigorous evaluation of performance.

Exxon Mobil, for example, has for many years required managers to set high performance standards and exits the bottom 5–10 per cent of performers. PepsiCo and GE have similar expectations. McKinsey has an 'up or out' system of advancement, which results in the majority of associates leaving the firm within three years.

8.6   Chapter summary

- Establish KRAs.
- Create the right balance between being task driven or too people focused.
- Performance management involves giving feedback in a way that is constructive and useful to the recipient.
- Poor performance should have been addressed throughout the year. The annual appraisal is the time for reviewing performance in a positive mode, and defining the way forward in the following year in light of changes within the organisation, and the person's aspirations for career development.

Chapter 9

Recognition and reward and the development of your team

> *This chapter will continue developing the theme of the last chapter, by reviewing the variety of ways that you might recognise, reward and develop members of your team following their appraisal. It will explore the normal range of incentivised pay reward schemes, and move on to other forms of reward – job enrichment, career development and personal skill development.*

9.1 Rewarding performance

As we have highlighted in the previous chapter, at the end of the appraisal process you will be looking towards rewarding performance appropriately and the ongoing development of the appraisee. This links back to the 'culture' of the organisation described in detail in Chapter 4. People observe what is discussed, recognised and rewarded, what is punished, what is tolerated and also ignored. So it is up to you to make sure that recognition and reward is pertinent and underpins the environment you wish to create.

First of all, importantly, you need to consider whether the appraisee has achieved results in their current role. If they have been offered training and development and appropriate support and have not delivered against an agreed standard, then you are unlikely to consider rewarding present performance or believe they are able to cope with more. Promotion and new opportunities may – rightly – elude them until performance improves.

For those who have achieved satisfactory results or exceeded expectations, then there are the usual incentives that can be used – performance-related pay, bonuses, and sometimes inclusion in an enhanced health plan or pension scheme, but these will be controlled by your organisation's policies rather than by you.

As an aside, some organisations such as Sterling Commerce offer a Flexible Options Programme. This programme allows staff members to create their own individually-tailored benefits package and also offers them the opportunity of realising additional cash. When established in the 1990s this was a leading-edge approach to benefit provision. While separate from the appraisal scheme, it offers staff the opportunity each year to alter their take-up of annual vacation, company pension and, if eligible, company car. By reducing the value of any benefits included in the programme, staff are able to realise cash in lieu.

Our experience of reward suggests that organisations must establish a more flexible approach to pay management in terms of each business and each individual. This means that pay and bonus awards should favour high performers and that the responsibility for pay and bonus decisions should rest with line managers. Pay systems – like all systems – should be simplified and transparent. This means, in the instance of the team leader, that both you and the team member have a complete understanding of their total reward package.

A best practice approach to reward says that:

- A 'one-size-fits-all' approach must go.
- Pay systems should be open, predictable and designed to demonstrate links to performance.
- They should be designed to recognise individual, team and group performance.
- They should reward according to a position's ability to create value.
- Managers and team leaders should become capable of the responsibility for managing reward.

Case study: 'Big four' UK bank[1]

Niall Foster worked as an interim director at one of the 'big four' banks in 2000. The results of a staff attitude survey showed that less than 20 per cent of 550 staff in his department considered that they used any other method of reward and recognition except money to encourage good performance. A research project was commissioned to determine how the department could rectify this, by identifying other ways of recognising and rewarding people. Forty-eight people across the department were interviewed and the findings were presented to management.* The documented options were discussed and the decision was taken to prepare an implementation proposal document outlining the chosen reward and recognition scheme and benchmarking criteria in readiness for launch in 2001.

*Findings from research

- Recognition should be about receiving thanks/praise for a job well done.
- Recognition didn't happen enough and did not generally extend beyond the immediate team/line manager.
- Existing communication methods did not support recognition, e.g. department magazine, a 'brag' board, team meetings, etc.
- People perceived there were behaviours attached to recognition such as team working, sharing best practice, etc.

- Reward is a personal issue – a one-'prize'-fits-all approach would not be satisfactory.
- Performance management was not deemed as motivational on an ongoing basis.
- Reward and recognition should be more timely and should be related to specific incidents.

Recognition and award scheme

The department created benchmark criteria based on the Bank's 'leadership and personal imperatives' to allow synergy with their performance development (appraisal) scheme and other personal development initiatives. A Recognition and Award Committee was created to discuss all nominations received and decide on the winning nomination. All nominees would receive a congratulations card. A monthly winner received an award up to a specified value (vouchers/tokens/surprise gift). The monthly awards were designed to recognise a *specific achievement during the previous month only*. The winner was given a choice about how to receive the award. Nominations for annual awards were for consistent demonstration of leadership and personal imperatives behaviour throughout the year. There were two runners-up.

The above shows that people want recognition. They want to be noticed. They want to be heard. They want to be listened to. Money helps but personal recognition or attention always wins.

So, moving on, let's look at what you *can* control to use as part of the appraisal reward system:

- job enrichment
- career development
- personal skill development

9.2 Job enrichment

Job enrichment is the addition to a job or task that increases the amount of control or responsibility a person has over their workload. It is a vertical expansion of the job as opposed to horizontal expansion, which is known as job enlargement.

Job enrichment has an important supportive and motivational role. Who does not come to work each day and want to enjoy him or herself? Does your organisation's culture support job enrichment? To us, this refers to people management and the team leader's ability to motivate and engage at a personal level with all who report to him/her. So what does this involve? For the team leader it is all about four essentials:

1. delegation – and/or simply asking team members for their ideas
2. listening – which makes them feel valued and a participant rather than a bystander
3. praising and complimenting rather than only focussing on the negatives
4. agreeing SMART objectives with team members

Job enrichment as a team-leader activity includes the following three steps:

1. turning a team member's effort into performance
2. linking their performance directly to reward
3. making sure the person concerned wants the reward.

1.   Turning team members' effort into performance

- Ensure that objectives are well defined and understood by all team members. The overall corporate mission statement should be communicated to all. Each individual's goals should also be clear. Every team member should know exactly how s/he fits into the overall process and be aware of how important their contributions are to the organisation, the team and its customers.
- Provide adequate resources for each team member to perform well. This includes support functions like information technology, communication technology and personnel training and development.
- Creating a supportive corporate culture. This includes peer support networks, supportive management and removing elements that foster mistrust and politicking.
- Free flow of information. Eliminate secrecy.
- Provide enough freedom to facilitate job excellence. Encourage and reward employee initiative. Flexitime or compressed hours could be offered.
- Provide adequate recognition, appreciation, and other motivators.
- Provide skill improvement opportunities. This could include paid education at universities or on-the-job training.
- Provide job variety. This can be done by job-sharing or job-rotation programmes.
- It may be necessary to re-engineer the job process. This could involve redesigning the physical facility, redesigning processes, changing technologies, simplification of procedures, elimination of repetitiveness or redesigning authority structures.

2.   Link team members' performance directly to reward

- Clear definition of the reward is a must.
- Explanation of the link between performance and reward is important.
- Make sure the appraisee gets the right reward if they perform well.
- If reward is not given, an explanation is needed.

3. **Make sure the person wants the reward. How do you find out?**

- Ask them.
- Use internal surveys to gather a fair cross-section of input.

9.3   Increasing job satisfaction

Most of us want interesting, challenging jobs where we feel that we can make a real difference to other people's lives and stretch our own capabilities. As it is for us, so it is for the people who work with us. So why are so many jobs boring and monotonous? And what can you do to make the work your team members undertake more satisfying?

The psychologist Frederick Herzberg, in his 1968 article 'One more time: how do you motivate employees?',[2] talked about enhancing individual jobs to make the responsibilities more rewarding and inspiring for the people who do them.

With job enrichment, you expand the task. You provide more stimulating and interesting work that adds variety and challenge to a person's daily routine. This increases the depth of the job and allows people to have more control over their work.

> Before you look at ways to enrich the jobs in your workplace, you need to have as your foundation a good, fair work environment. If there are fundamental flaws – in the way people are compensated, their working conditions, their supervision, the expectations placed upon them or the way they're treated – then those problems should be fixed first. If they are not resolved, any other attempts to increase satisfaction are likely to be sterile.

9.3.1   Designing jobs that motivate

There are five factors of job design that typically contribute to people's enjoyment of a job:

- **Skill variety** – Increasing the number of skills that individuals use while performing work.
- **Task identity** – Enabling people to perform a job from start to finish.
- **Task significance** – Providing work that has a direct impact on the organisation or its stakeholders.
- **Autonomy** – Increasing the degree of decision making and the freedom to choose how and when work is done.
- **Feedback** – Increasing the amount of recognition for doing a job well, and communicating the results of people's work.

Job enrichment addresses these factors by enhancing the job's core dimensions and increasing the person/people concerned sense of fulfilment.

9.3.2   Job enrichment options

The central focus of job enrichment is giving people more control over their work (lack of control is a key cause of stress, and therefore of unhappiness.) Where possible, allow them to take on tasks that are typically done by more experienced members of the team or by you as the team leader. This means that they have more influence over planning, executing and evaluating the jobs they do. We will be moving on to delegation in more detail in the next chapter.

Here are some strategies you can use to enrich jobs in your workplace:

- **Job rotation**
  This gives team members the opportunity to use a variety of skills and perform different kinds of work. It can be motivating, especially for people in a role that is very repetitive or that focuses on only one or two skills.
- **Combine tasks**
  Combine work activities to provide a more challenging and complex work assignment. This can significantly increase 'task identity' because people see a job through from start to finish. It allows team members to use a wider variety of skills, which can make the work seem more meaningful and important. For example, you could convert an assembly-line process in which each person does one task, into a process in which one person assembles a whole unit. You can apply this model wherever you have people or groups that typically perform only one part of an overall process. Consider expanding their roles to give them responsibility for the entire process, or for a more substantial part of that process.

These forms of job enrichment can be tricky because they may provide increased motivation at the expense of decreased productivity. When you have new people performing tasks, you may have to deal with issues of training, efficiency and performance. You must carefully weigh the benefits against the costs.

- **Create autonomous work teams**
  This is job enrichment at the group level. Set a goal for a team, then make team members free to determine work assignments, schedules, rest breaks, evaluation parameters and the like. You may even give them influence over choosing their own team members. With this method, you'll significantly cut back on team supervision, and people will gain leadership and management skills.

Job enrichment provides many opportunities for team members' development. You'll give them an opportunity to participate in how their work gets done, and they'll most likely enjoy an increased sense of personal responsibility for their tasks.

Don't just accept these points wholesale – they'll work in some situations and not in others. Apply them sensibly and in a way that is aligned with the realities of your workplace and your organisation's mission.

9.3.3 Implementing job enrichment

Step one

Find out where a person is dissatisfied in their current work assignments. There's little point in enriching jobs and changing the work environment if you're enriching the wrong jobs and making the wrong changes. Don't make the mistake of presuming that you know what a person wants. Ask them and use that information to review what you might be able to offer.

Step two

Consider which job enrichment options you can provide, in light of the other skill sets you have in the team. You don't need to drastically redesign your entire work process. The way that you design the enriched job must strike a balance between operational needs and job satisfaction.

Step three

Think through the changes you can make in light of the balance of workload of the team. If you're making significant changes, let everyone know what you're doing and why. Remember to monitor your efforts and regularly evaluate the effectiveness of what you're providing.

9.3.4 Key points

Job enrichment is a fundamental part of attracting, motivating and retaining talented people, particularly where work is repetitive or boring. To do it well, you need a great match between the way your jobs are designed and the skills and interests of the team members working for you.

When your work assignments reflect a good level of skill variety, task identity, task significance, autonomy and feedback, members of your team are likely be much more satisfied, more motivated and less stressed.

Your responsibility is to figure out which combination of enrichment options will lead to increased performance and productivity, bearing in mind the need for a fair and balanced workload for all in the team.

9.4 Career development

Organisations can vary in terms of the approach they adopt for career development of staff. Vodafone, for example, use their own competency model, a performance management process and development options guide. As a blue chip organisation they have the opportunity to have this robust process in place. The organisation's competencies framework forms the platform on which career management is built, and underpins all their people strategies from reward through to career and personal development. You will need to check with your HR department or line manager to find out if your organisation has a competency framework and career development guide and in turn discover the protocol to aid you or your team members toward advancement.

In practical terms, how can you help appraisees to think through how they might want to progress their careers? You need to get them to move on from purely technical competencies to other factors that are of significance, which will be assessed once they wish to progress. For example get them to:

- **Assess their broader skills**
  This could be anything from communication to problem solving and managing people or projects.
- **Assess their work values**
  Here they could consider factors such as having a:
  - strong need to achieve
  - need for a high salary
  - high job satisfaction requirements
  - liking for doing something 'worthwhile'
  - desire to be creative, travel or be independent
  - do they want to continue working as part of your team, or any team for that matter!
- **Assess their personal characteristics**
  Are they a risk taker, an innovator or able to work under pressure, and how do such characteristics affect their work situation? Another consideration is to discuss if they prefer working independently on a particular aspect of any assignment. If they can add value through their contribution to the team effort but shy away from group participation, then the people management route is not for them.
- **Assess their non-work characteristics**
  Factors like family commitments, where they want to live and their attitude to time spent away from home.
- **Match their analysis to the market demands**
  Consider realistically how well their overall capabilities and characteristics fit market opportunities.

With this clear, you can then jointly set defined objectives. The old adage, 'if you don't know where you're going, any road will do', is nowhere more true.

The next step is to encourage them to aim high. They can always look to trade down, but they should not miss out because they did not attempt it.

9.5   Personal skill development

There are many ways of learning, developing and growing (see Figure 9.1). Personal development should not be considered as something that only happens on a training course. In reality some of the most effective ways of learning can come from undertaking day-to-day activities.

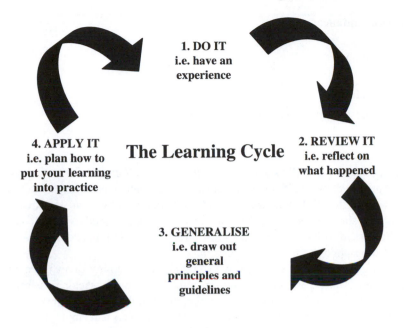

Figure 9.1   Personal skills development and the learning cycle

Your team members' training need requirements will result from the appraisal and coaching process, and other ad hoc performance-related issues.

So how should you go about developing others?

9.5.1   Learning and development resources

There are a variety of learning and development resources that can be used.

Self-learning through the use of:

- knowledge portals
- intranet/Internet
- distance learning/computer-based training/CD-roms and DVDs

'Experiential' learning and development:

- on-the-job training
- 'sitting with Nelly'
- undertaking specific tasks and projects
- job rotation
- secondment and attachment
- organising a best-practice benchmarking visit

Through guidance:

- coaching
- mentoring

Formal training:

- in-house or public course training programmes

> According to the CIPD's annual *Learning and Development Survey* published in April 2008, over half of the learning and development managers (57 per cent) surveyed offered eLearning as part of their training provision. However, the findings showed there still continued to be doubts about its effectiveness. When asked to list the top three most effective training practices, only 7 per cent of respondents mentioned eLearning.
>
> Currently, larger organisations with 5,000 or more employees are more likely to use eLearning, particularly in the government and public sector, possibly due to the government's endorsements of eLearning in the Leitch Report.[3]
>
> Despite eLearning apprehension from both employers and employees, nearly half of the respondents agreed it had been the most important development in training in the last few years, and almost one-third said it was likely that in the next three years between 25 and 50 per cent of all training would be delivered remotely.

There should be some form of learning resource library either in your department or in the HR/training department. If there is not, create one yourself to include the latest range of publications in your industry sector, and best practice case studies from a variety of industry sectors around the world. Let's face it, people have similar motivational/problem-solving issues regardless of the sector they are in, and we can all learn a lot from other organisations' experiences.

Below is some further information about each category.

9.5.2   Self-learning

If you wish to simply source information knowledge portals, the intranet/Internet and distance learning all have their place.

If you require a more in-depth exploration of a topic, however, we would suggest you adopt a blended learning approach. This could involve computer-based training, the use of CD-roms/DVDs and also, importantly, getting together with others for some form of workshop on a regular basis. When Pat worked in higher education she researched potential delegates to discover what forms of training and development worked for them. Consistently, the feedback was that people found it too hard to work on their own for any protracted period of time. An exchange of ideas and learning from others was as much a part of the learning process as the material supplied.

9.5.3 Experiential learning and development

As mentioned earlier, learning can be informal, for example, challenging your reports on a regular basis by asking, 'how/what/why ... did you tackle x assignment in a particular way, and what have you learnt from the experience?'

Moving on to a slightly more formal approach, if you undertake 'on the job' training, you need to be aware of how people absorb information. Are they auditory, visual or kinaesthetic learners? People use their senses in different ways to take in information and learn (see Chapter 12). Some are *auditory*-type learners: they learn best by listening to knowledge told to them. Some are *visual*-type learners: they learn best by seeing pictures and symbols (or plain words). Others are *tactile*-type: they learn best when they can touch and examine objects for themselves. Some combine two or more senses depending on what they are learning.

During a development session, people have a limited attention span while sitting in one place absorbing information. If training or guidance is being passed to an individual, they will be able to concentrate for a maximum time frame of only forty-five minutes. This is if just one methodology is being used, say, delivering the verbal part of a training session, even with the use of visual aids. People need to practise what they have learnt through the use of a case study, exercise or syndicate work. Training on the job should always have a practical 'hands on' application applied once core information has been supplied.

Remember as well that if a person does not use the information that they have learnt during a development session they will usually forget 75 per cent of it within three weeks to one month.

A final key point to remember is that you are far more motivated to learn if you need to achieve an outcome of some kind, rather than just absorbing information for information's sake.

Looking outside your team, secondments and attachments to other departments are also a good way of developing people's skills and competencies. If a specific secondment or attachment is identified during an appraisal discussion, you will probably need to escalate a request to a more senior level, for example to an executive director who is ultimately responsible for the individual prior to any commitments being made. This executive director will then normally agree the best way of discussing the attachment with the other parties who need to be involved, i.e. another executive director or HR.

Case study[4]

Dupont's early career process explicitly moves engineers across different sites for developmental purposes, and Dow Chemical's website similarly points out to career prospects that, 'we offer employees opportunities to work as part of a global team on projects with a global scope far more often and earlier in your careers than other employers'.

9.5.4   *Learning through guidance*

Coaching and mentoring

First of all, let's look at the difference between coaching and mentoring.

Coaching is focused on improving an individual's performance against a task, providing feedback on their strengths and weaknesses and helping them improve their performance. Mentoring, on the other hand, is focused on a longer-term, deeper relationship that is aimed at helping an individual to learn about appropriate behaviour in a particular organisation, how to navigate the political climate, and the cultural dynamics of 'how things work around here'.

You might have a mentor yourself (more about this in Chapter 15: Planning for the future). In turn, a member of your team might wish to seek a mentor within the organisation. In some organisations this process is centralised and formalised; a person is allocated or there can be a request for one, making this a regular development technique. Usually a mentor is not person's team leader or immediate manager, so you could be asked to facilitate finding a potential candidate. What starts as an apparently one-off agreement may act as a catalyst and lead to a more substantial arrangement.

So let's look at coaching in more detail.

Principles of successful coaching

Coaching should:

- be one to one, and in privacy if possible
- start from where the person actually is in performance terms, not where they ought to be
- build on personal strengths and aim to remedy weaknesses
- be regular and involve constant feedback
- be person driven rather than task driven
- be a joint process

The effectiveness of coaching depends on the skills of you as a coach and the receptivity of the person being coached. Success is therefore more likely in conditions where:

- there is clarity about expectations.
- success criteria is clearly established
- clear and regular feedback is given.
- There is agreement on the resources required to achieve targets.
- there is understanding of the work context.
- rapport and trust have been established.

How to conduct a coaching session

Team leaders must think of their team members in terms of their (future) potential, not just their (past) performance. For you to have the capacity to coach is not only beneficial to the recipient, but also to you personally in terms of your career progression.

It is a fact that coaching competencies, based on observed behaviour and 360-degree feedback, are heavily weighted in the selection of managers as they progress to a more senior level within an organisation.

Coaching is about unlocking potential and getting people to think for themselves, so that they can maximise their own performance. No two human bodies or minds are the same. How can I tell you to use yours to your best ability? Only you can discover how, with *awareness*.

Our potential is realised by optimising our own individuality and uniqueness, never by moulding them to another's opinion of what constitutes best practice. To 'tell' denies or negates another's intelligence. To 'ask' honours it. Telling or asking closed questions saves people from having to think. Asking open questions causes them to think for themselves.

Coaching questions compel attention for an answer, focus attention for precision and create a feedback loop. Instructing does none of these.

When I *want* to, I perform better than when I *have* to. I *want* to for me; I *have* to for you. Self-motivation is a matter of choice. Coaching offers personal control. A primary cause of stress in the workplace is a lack of personal control.

Here are some useful tips on the key stages of running a successful coaching session:

- State the purpose and importance of the session. You need to be clear in your own mind about what you are trying to achieve before you begin.
- Establish and agree desired outcomes.
- Seek to establish improvements.
- Give objective feedback.
- Discuss options.
- Agree objectives and development options. Remember to make sure they are SMART.
- Agree follow-up meetings.

John Knapton is a leading business coach in Northern Ireland who specialises in executive coaching and mentoring. He shared with us the following case study.

Case study: Coaching in action

Supporting employees to achieve their goals means constant open, honest communication and immediate feedback, ensuring that the employee is 'right' for the role and providing the appropriate training to make certain that the individual is fully equipped with the knowledge, skills and mind-set to succeed. There are, however, times when some individuals need more support than others; good examples of this are when a change in position occurs, when they are performing below their ability or when they have been assigned the wrong role.

One such occasion arose recently with a key individual in a marketing team I was working with. Martin, an experienced product manager with responsibility for a significant proportion of the company's revenue generation, was suffering badly from a lack of credibility with the new management team. Martin demonstrated excellent technical, product and application knowledge, but his product lines (although profitable) had stagnated in growth. His lack of forward strategic planning and attention to financial details, coupled with his style of communication when presenting to the management team, dramatically diminished his standing and his ability to drive the changes needed to revive the growth in his product portfolio.

The management team were faltering with regard to Martin's effectiveness and the viability of his contribution to what was viewed as a growth market.

I believed that Martin could benefit from some individual coaching, utilizing his natural enthusiasm, knowledge and ambitions to identify his own internal barriers to success, to become self-aware and to take ownership of the changes that would be required for him to succeed and regain credibility within the team.

The coach should have a unique relationship with the coachee, supporting him through the process of change. If the coach does not believe in the coachee's ability to achieve his stated goals, then a problem exists and it is unlikely that the coachee will be successful. It is therefore paramount for any coach to believe in the success of the coachee and his abilities (albeit abilities that are not currently utilised to best effect) and to build a positive and open rapport with the coachee from the outset. By empathising (not necessarily agreeing) with the client, by mirroring his body language and by modifying mood, voice tone and the structure of communication, the coach can quickly build rapport with the client, assisting him in 'being at ease' and allowing the coachee to open up fully to the coach. Having established this early in the initial coaching session, Martin communicated his openness to learning and change and to coaching as a means to decrease his stress level while increasing his relative effectiveness and standing in the eyes of his peers and management.

The level of belief in himself (the degree to which the coachee believes he can reach his personal goal) and the general attitude the coachee displays can all be worked through during coaching conversations in order to explore the possibilities that change can offer. Also important in any coaching session is the end, the point at which the current coaching relationship can be dissolved. By using SMART objectives and by laying these out for both the whole of the coaching experience and for each session, short- and long-term wins can be recognized by the coachee. By recognising progress towards his final goal, a sense of continuous achievement can further add motivation and aid the development process.

I also found it useful during my coaching sessions to objectively explore the coachee's attitudes and belief systems. By engaging unemotionally, I was able to tease out the real truth behind attitude and actions. I discovered, for example, that my initial interpretation of some of Martin's actions and what I assumed to be his underpinning beliefs were incorrect. Using a non-directional coaching approach, I was able to understand the real drivers behind his actions and subsequently support him better through the change process. His propensity to use repetitive phrases during presentations turned out to be neither an inferiority complex nor a nervous disposition as I had originally thought; rather, it was a displacement activity to allow himself time to think, his belief being that 'silence' would be interpreted by his audience as weakness. Misplaced assumptions by Martin in the first instance, therefore, resulted in the activity evolving into a debilitating habit, the key component that adversely affected his credibility – (the non-verbal negative feedback compounding his degenerative self-confidence and manifesting itself in his ability to think innovatively) – the opposite effect of what he actually intended!

From a coaching standpoint, additional time must be taken to understand situations from others' perspectives. The alternative is potential failure in achieving the coachee's objectives because of personal subjectivity at best, or a biased, emotional involvement that leads to a degeneration of the coaching relationship at worst. When perspective on particular problems became limited, inhibiting real innovation, I found it very useful to ask Martin to 'step outside' his own mind and look at the problem from someone else's perspective, his line manager, his peers, his wife's, the CEO's. This usually created new perceptions and often led to a greater self-confidence. It was interesting to note that Martin could be very linear in his thinking, giving the initial impression of limited strategic thinking; however, thinking as a third person, he was able to create new scenarios and demonstrate significant creativity. This ability was always present; it just needed unlocking.

Throughout the coaching process it is important to ensure that ideas, actions and development plans are the coachee's own. When mingling mentoring or interpretation of the organisation's development needs within a coaching session (a perfectly acceptable paradigm, I believe), the coach should always ask permission, presenting new thinking and ideas as 'additional options' to the coachee's own list. In operating in this manner, the coachee retains full ownership and responsibility for his vision while considering all of the available options and potential outcomes on both a personal and organisational level. Questions beginning, 'what else . . . ' or 'what if . . . ' can be very strong stimuli in creating a vision and articulating clear objectives for the coachee and subsequently for the organisation.

Coaching is a very powerful tool, only relatively recently becoming recognised as an essential part of personal development plans and a significant plus in releasing individual potential, creating 'A' players who benefit organizations at every level. In Martin's case, the outcome was no less than extraordinary. Just four one-hour sessions and a brief, 10-minute confidence-building chat before his next presentation to the senior management team resulted in a crisp and clear message of new thinking portrayed and literally gasps of 'wow' from the meeting attendees. They could not believe what they were seeing and hearing, having already 'given up' on Martin. Real, engaged discussion ensued, together with a very positive anticipation for the next product review meeting.

So to establish the key learning points from this case:

* Make sure you give time and space for the person you are coaching to arrive at their own conclusions.
* Remain as objective and non-judgemental as possible and ask 'what if . . .' and 'what else . . .' questions to prompt lateral thinking and allow the person to visualise their response options.
* Make sure as you progress that there is a clear sense of direction to your conversations and SMART objectives are developed by the coachee to take things forward.
* Believe in the coachee's ability to achieve their stated goals.

9.5.5   Formal training

You and your team members can keep up to date with professional knowledge and develop pertinent skill sets through attending training programmes, be they in-house or public courses. Being part of a professional body or association will also provide a further range of ideas and activities to support professional development. Networking with other similar professionals at conferences can also help in career development terms.

As you can see there are a broad range of options that can be called upon to develop your team.

Creating a learning environment is essential for organisations to survive. So often a company has been set up and been successful as a result of a few core products or services. They gain market share, but in an increasingly competitive environment new entrants can rapidly enter the market and undercut on price or range of added-value services. The buying behaviour of customers can also change. This means staff at all levels need to challenge the status quo, ask questions, be customer-focused and keen to learn in order for your organisation to be one step ahead of the competition.

Case study

IBM publicise an extensive education programme on their website. Staff have 100 per cent tuition fee reimbursement and a comprehensive technical and professional curriculum.

ANZ Bank of Australia has implemented a six-month sabbatical programme to engage senior technical staff with important vendors. Their goal is to facilitate the continued development of technical skills and to stay in touch with new developments.

9.6 Chapter summary

- Recognition and reward can be offered in a variety of ways – it is not always down to money.
- Job enrichment through expanding tasks and normal work activities can enhance a person's life.
- Look to enhance the career development of your team members by discussing with them optional activities that can develop their skill sets.
- Choose to train and develop team members from a plethora of activities, as and when required.

Chapter 10

Team enablement/empowerment and delegation

Team enablement/empowerment and delegation require self-discipline and altruism. These skills should exist with a great team leader. This chapter will explain how your skills in leadership can promote motivation, loyalty and self-esteem from team members that in turn will lead to their self-development.

Chinese proverb

TELL ME and I will forget
SHOW ME and I may remember
INVOLVE ME and I will understand

10.1 Enabling and empowerment

Enabling/empowering your team members is best practice. We constantly find that there is no true understanding or agreement in organisations of what enablement/ empowerment actually means. All too often team members feel that management want 'things yesterday' without consideration for current workload or work priorities and with no consideration for internal and external customer needs and requirements. Team members feel 'dumped on' without support or guidance. They flounder because they don't have adequate resources, and don't know where to go to gather information.

This chapter will look at what it takes to enable/empower the members of your team and how to delegate effectively to make it work. Team leadership and management in general is about delegation. If you don't delegate you can't motivate. As a team leader you will inevitably need to pass work to members of your team. This will not only reduce your own workload, but will also give others the opportunity to develop their capabilities and broaden their skill sets. This can be highly motivational, as people can feel valued for the contribution that they are able to make. If people feel an enhanced sense of worth, they will have a greater sense of commitment to their working activities. When people are communicated with effectively, given skills, tools and responsibility, they thrive.

Prior to delegating, you need to be clear in your own mind as to the main objective of your job and how you should be spending your time, and see what tasks might be done by someone else. It could be that a member of your team has a good graphic eye (for example, do you always have to be the person that creates the PowerPoint visuals in your reports, or could one of your team do these?) An individual could also have more technical expertise than you at a particular phase of a project, and this will fast-track the overall delivery time of the assignment. Bearing in mind the need for continuous development of your team members' skill sets, there could often be occasions when you could get a junior team member to stand in for you, and by doing so, let them gain experience. Meetings often provide a good opportunity for this approach.

As we have discussed in previous chapters, the development of communication skills for staff is vital in terms of their being able to contribute constructively and intelligently during team meetings, quality circles and continuous improvement activities. If you are going to delegate, you also need to be sure that the person involved has the skills and experience to undertake the task. This could involve just a small amount of coaching to get them up to speed, or could take the form of more comprehensive 'just-in-time' training and development.

The enablement equation (Figure 10.1) gives a graphic illustration of what needs to be in place to give staff the fundamental knowledge and support they require to be proactive and to take forward an activity in an energetic and well informed way.

To enable/empower is to move from giving instructions as to how to do a particular task or activity, to giving the team or individual the power to act on their own initiative, to take responsibility as to how to go about tackling it, and to move forward to the implementation stage. It's an obvious idea: the people closest to the work understand

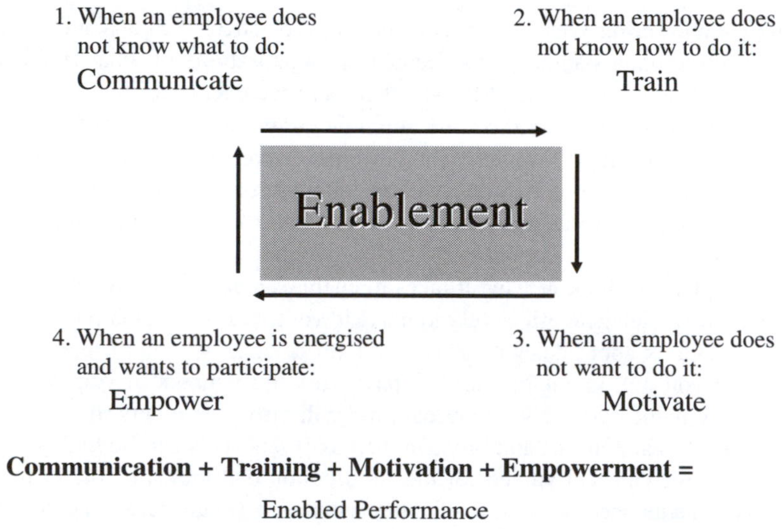

1. When an employee does
   not know what to do:
   Communicate

2. When an employee does
   not know how to do it:
   Train

Enablement

4. When an employee is energised
   and wants to participate:
   Empower

3. When an employee does
   not want to do it:
   Motivate

Communication + Training + Motivation + Empowerment =
Enabled Performance

Figure 10.1   The enablement equation

it best; they are the experts and know what the customers want because they work with them all the time.

Empowering members of the team does not mean handing them company command and control. Ultimate decision-making authority must remain with you as a team leader and with management (plus shareholders) – but it should enable them to make and act on local decisions which influence their own work. This means that each team member and team must be given a defined arena of functional freedom within which they can exercise their knowledge and skills. Decisions or actions which potentially take an individual or group across their area boundary will, of course, be subject to team leader or management sanction and they will be aware of this. But inside the arena the team or individual must be free, that is enabled/empowered, to operate according to their own sanctions.

We cannot emphasis enough that empowering employees is not giving them command. If you are unsure of the capabilities of the person you are delegating to, ensure that they come back to you with their implementation proposal before they move forward. The key here is that you sign off/agree their plan before they proceed, because ultimately you always retain responsibility.

In the early 1990s a leading European consultancy firm lost the majority of its management team. Why? The management presented to the board a strategic implementation plan that had already cost a million US dollars. The board refused to accept their proposals. The staff who had created the plan felt demotivated and many left the company. So what had gone wrong in this scenario? The management team got carried away and forgot that there was an ultimate decision maker – the board of directors – who had the final say. They proceeded with creating the strategic implementation plan *without getting prior approval*. The end result was that the company not only lost key management staff, but also a million dollars.

Giving your team this power, albeit on a local scale, requires four things from you as a manager:

1. Confidence in your team, so that you manage *by exception.*
2. Trust in your team members to use their empowerment beneficially.
3. Continuous dialogue and feedback from you to reinforce the company culture and values.
4. Leadership from you that reinforces the company mission to provide first-class customer care.

The result of empowering your team is conscious and observant attention to detail and improvement. For example, Marriott International have had a long-term policy of empowering their staff, and this has encouraged and enabled them to take the initiative in a variety of small ways, all for the benefit of the customer.

*   A guest at a Marriott hotel lost some coins in a drinks vending machine, so he asked a cleaner who was passing how he could report the broken machine. Without hesitation the cleaner put his hand in his pocket, gave the guest a refund, and promised to report the problem immediately, which he did. His manager reimbursed the money.

- Two businessmen were to have a meeting over lunch at a Marriott hotel, even though they were not staying at the property. While walking to the hotel they were caught in a heavy downpour and arrived at the hotel very wet. Rather than sitting uncomfortably in the restaurant, staff took the initiative and arranged a guest room for their use, with a room service lunch and bathrobes. Their clothes were dried and pressed while they had lunch in private.[1]

> People do not get empowered from others: they empower themselves.
> People-based leaders establish conditions and contingencies to facilitate empowerment, but they do not give empowerment; they enable the release of empowerment from others. It is not about getting empowered, rather it is about feeling empowered.[2]

So, how do you go about creating an enabled/empowered environment where team members can take the initiative in this way? There are five well-known levels of decision making, most of which are modelled after Tannenbaum and Schmidt.[3]

- **Level 1** *Tell/directive*
  This could be a straightforward instruction, or the manager reviews the options and tells the team what he/she has decided.
- **Level 2** *Sell/input*
  The manager explains the rationale behind the decision, putting an emphasis on the benefits to the team.
- **Level 3** *Consult/dialogue*
  Background information is given to the team as well as the decision, and the team are invited to ask questions and have further discussion with the manager.
- **Level 4** *Participation/ownership/consensus*
  The manager invites discussion with the team from the word go, but still makes the final decision.
- **Level 5** *Delegate*
  The manager explains the problem or situation which needs to be resolved, defines the limits and timeframe involved and it is then up to the team to take the project forward to completion.

How you delegate depends very much on the experience and ability of the person involved or the maturity of your team, and there are going to be certain activities where it is not possible to enable/empower staff. This could be in relation to legislative requirements, financial control mechanisms, legal constraints, security or specific standards that need to be adhered to for quality control purposes.

   Looking at the different levels above, there are occasions when you simply tell a group or an individual what needs to happen – a straightforward instruction (level 1). It makes sense that the more people feel involved and included in the

Directing	Facilitating
• Tell team members what processes and tools to use	• Suggest processes and tools – let team members choose
• Provide answers to team members' questions	• Coach team members to find own answers
• Take direct actions to ensure team makes good decisions	• Help team make their own decisions by facilitating discussion process
• Direct team meetings and progress	• Observe team meetings and monitor progress
• Provide resources; link team members with the rest of the organisation	• Reinforce team members' ability to find their own resources, make their own links
• Intervene in team activities frequently to keep team on track	• Intervene only when team members are not making any progress
• Teach team members basic tools of project management and team building	• Teach team members advance tools; reinforce use of basic tools

Figure 10.2   Moving from a directive to a participative approach

decision-making process, the more they will feel motivated to think laterally and creatively and come up with ideas themselves as to how a task should be tackled and implemented.

In Figure 10.2 you see how you can move from a directive to a more participative approach, facilitating others to take things forward.

10.2   Prerequisites for successful delegation

Your attitude and support as a team leader during the delegation process is key to successful empowerment. Team members need:

- clearly stated objectives as to what they need to achieve and specific reasons why
- an exact description of the quality standards that need to be achieved
- a realistic time frame and access to budget should they require this
- guidelines in place as to the scope of empowerment that is being exercised, and where the limitations lie
- clarification of where the team member/team can source information or resources

- the authority they need, arrange in readiness
- discretionary powers to do the job
- the right to make mistakes and to use these mistakes to learn from (as we have already explored in a previous chapter)
- mentoring, coaching and constant feedback
- your assurance that you will represent their interests and keep the predators at bay
- to know what kind of progress reports are wanted, when, and how often
- for you to identify the conditions under which he or she should contact you for assistance

You have to bear in mind that initially, at least, it can take you more time to delegate than doing the activity yourself.

It is important not to keep checking up. Leave team members time to think and reach their own decisions. If someone comes to you for advice, try to encourage them to think through the problem themselves by asking them a series of open questions about the issue concerned.

Above all else, do not take credit for other people's work. Acknowledge their input; celebrate how they have tackled a particular task and the results they have achieved.

Thinking at an individual level, sometimes it is obvious who you should delegate a task to. But equally well you might have to think in more detail about the skills and competencies in your team.

10.3   How to utilise the diversity of skills and competencies in your team

Some very interesting work was carried out by R. Meredith Belbin at the Henley Administration Staff College over a period of several years.[4,5]

Belbin started out with a simple idea that different types of people interact in different ways. The research identified that there are a number of distinctive roles which people play in teams and that the ideal (successful and effective) team is made up of a mix of people who fulfil different roles.

An understanding of different team roles and contributions can help you and team members understand both the function of a team and their own role within it. It also helps everyone value the differing contributions made by colleagues, and helps you when you are planning to delegate an assignment.

It is important to recognise that any team is only as good as the sum of its members. Each member is different but important. Team members don't need to be all the same. In fact, a successful team needs a mix of people, personalities and skills to be successful. This is particularly true when dealing with the complex demands of an organisation and its internal customers. An individual may, and often does, exhibit strong tendencies towards multiple roles. In small teams an individual can certainly play more than one role.

In the matrix below you will see a brief resumé of the team types that were identified during the research project:

Type	Typical features	Possible contribution to the team & organisation
Chairperson	Calm, self-controlled, confident	• Discussion of needs • Agreeing shared objectives
Shaper	Highly-strung, outgoing, dynamic	• Pushing things through • Re-enthusing colleagues • Creating ideas
Plant	Individualistic, serious-minded, unorthodox	• Coming up with new ideas and approaches to give customers what they want • Creative and enabling use of systems
Monitor-evaluator	Sober, unemotional, prudent	• Offering objective feedback • Defusing arguments
Company worker	Conservative, dutiful, predictable	• Responding to needs • Getting things done, quietly and to a standard
Team worker	Socially oriented, rather mild, sensitive	• Communicating well with people • Getting details right • Working to satisfy others
Resource investigator	Extrovert, enthusiastic, curious, communicative	• Finding ways to provide what people want • Keeping up to date with customer requirements
Completer-finisher	Painstaking, orderly, conscientious, anxious	• Tying up the loose ends • Making sure everything happens as intended

So, using this behavioural tool, when you are thinking about delegating a particular task or activity to a member of your team or setting up a project, how can you best utilise your reports?

Taking a few of the characteristics identified above:

The term 'plant' was used by Belbin because this type of team member appeared to 'sit in the corner' and not interact a lot, rather like a house plant. **Plants** are creative, unorthodox and good at generating ideas. For in-depth problem solving or at the start of a new project they are an excellent resource to use. They often bear a strong resemblance to the caricature of the 'absent-minded professor', and are not always good at communicating ideas to others. You may need to tap into their creativity and draw out information from them by using the facilitating open questioning techniques described in the communication chapter.

The **resource investigator** will often have a great deal of enthusiasm at the begin ning of a project, be a good networker and tap into contacts outside the team to gather information to bring back for an assignment. When Pat first moved into consultancy, there was a researcher in her department who fulfilled this role to a 'tee'. She spent a proportion of her time doing classic research at the London Business School, gathering information for consultants. However, a great deal of the information she gathered was through chatting to staff in the coffee room, getting 'grapevine' information to pass around the building about the linkage between the different assignments staff were undertaking, and in turn where there might be cross-selling business opportunities.

Another two roles worth a mention are the **chairperson** and **shaper**. Both are leadership roles, and they can be complementary. The chairperson observes the team and knows the strengths and weaknesses of each person in it, whereas the shaper will often be the one who challenges and stimulates discussion – questioning approaches, etc. Too many shapers in the team, according to Belbin, can lead to conflict, aggravation and in-fighting.

Further interesting research has been undertaken by Sallie M. Henry and K. Todd Stevens in the Department of Computer Science, Virginia Tech., where they explored the use of Belbin's leadership role to improve team effectiveness.[6]

They conducted a controlled experiment with senior software engineering students, looking at Belbin's roles from a performance, productivity and team viability point of view. In a laboratory setting, a number of teams were formed each containing a single leader, and others were formed that had no leaders or multiple leaders. The conclusion of the experiment was that a single leader in a team improved a team performance over teams having multiple or no leaders. The mean time-to-completion of the leaderless team was significantly greater than the team with leaders.

And finally, the **completer-finisher**. For any major project it is important to have a person who is concerned about accuracy, making sure the layout of documents is perfect, who is happy to check and re-check what is to be delivered and ensure that the project has achieved its objective.

In reality, it is not always possible to set up an assignment/project team with exactly the right mix of individuals. If, however, you think each of the roles represents a team process, then you need to make sure that these processes all occur during the activity.

So let's look at a couple of case studies of what delegation can mean in practice.

Exercise

You are a manager. You meet members of your team informally, discussing with them the reduction of the department's costs by 6 per cent this year. Later in the day you attend a management meeting where it is agreed that there is a need to reduce your department's costs by 12 per cent. One of your ideas is to reduce the number of regional offices from four to three.

You arrange a formal meeting with your team to let them know of this decision.

How do you go about communicating this and motivate your team members at the same time? What are the challenges here?

- It is your role as a manager to communicate management decisions and delegate.
- Where does enablement/empowerment start and finish?
- How would you prepare to communicate this decision?

Task 1 *Communicate the management's decision*
For example, 'I have come from the management meeting, and we have decided to reduce department costs by 12 per cent this fiscal year.'

Task 2 *Communicate the specific reasons why 12 per cent was selected*
Ensure that all the management team members communicate the same reasons for the decisions made. This will reduce organisational rumours dramatically.

Task 3 *Ask the team members for their ideas to make this 12 per cent reduction in costs*

Give them time to prepare, by agreeing a date when you will meet to discuss their suggestions.

What are the points to note here?

- The management's decision cannot change. This is not negotiable.
- Giving reasons allows others to understand the rationale behind the decision. They may not agree with the decision, but if they understand why it has been taken this can help.
- Forget your ideas about reducing the regional offices from four to three. Why? Because the team members may come up with better ideas.
- Telling your staff what to do does not lead to their development.
- Agreeing a specific date to meet and discuss their ideas is true enablement. They come with ideas and you jointly agree the next steps.

For several years Pat Wellington worked collaboratively on a project with Cynthia Alexander (Elliott), the Head of Training & Development of a London borough, to deliver part of an MBA suite of programmes for a targeted sector in the council.

Case study

I (Cynthia Alexander Elliott) worked as the head of Training & Development in a London local authority, which was undergoing major change to improve organisational performance and customer satisfaction. During this period I experienced increasing demands on me and the department to create solutions and initiatives, to develop key employee competencies, to transform the culture and to improve performance.

At the time the fundamental requirements were to improve management effectiveness, develop leadership capacity and improve customer service.

After reviewing the organisation's priorities, I realised that I would have to work more strategically to reposition my department to deliver the expected outcomes.

My personal performance targets included implementing an effective customer service programme, assisting with the development and implementation of management and leadership development programmes, while assisting the organisation to implement relevant processes and practices to achieve the Investors in People standard (IiP) over a two-year period.

This was clearly a challenging period and my initial response was to assess and re-prioritise my work and activities, to enable me to focus on urgent and important things. Once I completed my analysis, it became apparent that I had to let certain things go. I considered various strategies to assist in managing the competing priorities, and decided to use delegation as one of my key approaches, as it would enable me to achieve a win–win situation, for me, the team and the organisation.

Interestingly, making the decision was much easier than implementing it. Once I had defined the tasks, I realised that the activities that were taking up a large proportion of my time were the ones I really enjoyed. As a particular example, I enjoyed working closely with suppliers who were delivering training sessions for me, but realised that with my new role I had to pull back to just being involved with programme product development and quarterly quality control monitoring, rather than being so closely involved with operational issues.

This was a real dilemma for me, I really struggled to let go, for key reasons which included: tasks might not be done as effectively by someone else; I would lose control; it would take too much time to train someone else, etc. Thankfully, after a certain amount of deliberation I was able to put things into context and see the bigger picture and the potential benefits. I started to let go gradually, by assessing the skills and abilities of staff before delegating tasks. I set clear performance targets and, most importantly, used coaching and shadowing to enable them to learn and develop their skills and confidence.

I also learnt a number of things on the way, before I became fully competent at delegating effectively. These included the need to explore possible risk factors as part of delegation process, to minimise risk, to delegate activities in stages until the delegate is confident, then let go and manage by exception.

Overall I learnt that for organisational and personal development we have to come out of our comfort zone and let go of the things we like doing, to enable us to become more effective. The key benefits were that I was able to achieve my targets in assisting the organisation to improve performance and customer satisfaction, while developing my strategic and leadership skills. Team members felt empowered, motivation and performance improved and my delegation approach was cascaded to junior staff, to facilitate further learning and development. The organisation achieved the IiP standard, my department has grown through reorganisation and team members have been promoted as a result of undertaking higher-level duties through delegation.

10.4   Handling resistance

Taking a 'devil's advocate' view on delegation/empowerment, you could be faced with resistance from a member(s) of the team who simply is not interested in expanding their horizons and taking on new challenges. This could be caused by a variety of factors – lack of self-confidence in being able to do the task, an 'I am not paid to do that job' attitude, 'I am too busy already', etc.

It is true that a few people lack ambition, and can never be motivated to accept further responsibilities or vary their workload. Possibly in the past the process of delegation has been handled badly, and you need to persevere and ensure that the way you delegate is supportive to the individual concerned. If they still resist, you could try teaming them up with a colleague who has a more open-minded approach and who is willing to develop their working activities, and see if this energises them to contribute more. There is no easy answer to this, and all of us as managers have had to face this dilemma. Fortunately, this is the exception rather than the rule. See Chapter 8 on how to handle underperforming team members.

'I have often found that small wins, small projects, small differences often make huge differences.'

Rosabeth Moss Kanter[7]

Having looked at the process of delegating a task or a project to an individual, let's move on to the considerations involved in delegating to a team.

10.5   Considerations in delegating a project

10.5.1   Set a clear agenda

Team members need a clear sense of direction *quickly*. How else can they be effective as a team? Clear priorities help team members figure out how to spend their time. The action plan sets out the agenda with crystal-clear tactical objectives giving the team laser-like focus. Alignment of effort depends on your ability to orient the team and orchestrate a coordinated effort. Map out new priorities, keep them simple and tie them to a specific timetable. Set short-term goals that the team can achieve quickly. Potential resistance can be defused when your instructions are unequivocal and easily understood. Make known your commitment to them and their commitment to achieving the goals. Tell them at the outset that they can expect some mid-course corrections. The agenda will have to be adapted as the situation demands it. But *always* keep it clear and communicate it constantly.

Of course, key team members can contribute to designing the team's priorities and objectives. You must consider their input. The more they can shape the agenda, the more buy-in and commitment they'll show. Plus, their ideas might dramatically improve your sense of priorities. In the final analysis, though, *you* remain accountable.

10.5.2   Ensure roles and responsibilities are understood

Ensure everyone knows what's expected of him or her. Don't leave people to figure things out on their own. Get rid of rule ambiguity. Nail down every team member's responsibilities with clarity, precision and attention to detail.

There must be no question regarding where one job stops and the next one starts. Do not blur the responsibilities each team member is supposed to shoulder. Figure out precisely what needs to be done, who is going to do which part of it, and then communicate your plan. Give every team member a brief job description. State your expectations regarding standards of performance. Describe the chain of command in the team. Outline each person's spending limits, decision-making authority and reporting requirements. Everyone will be best served if you put this information down in writing.

Check to make sure that each team member understands the team's (whole) set-up and how it fits together. Be careful to avoid job overlap, since that feeds power struggles, wastes resources and frustrates everyone involved. When explaining to people what to do, also specify what they should *not* do. Differentiate between crucial tasks and peripheral, low-priority activities. Spell out what needs to be accomplished in each position and for what the person will be held most accountable. Once you have done this, pay attention to what team members are doing. Keep everyone on track and if you see something going wrong, fix it immediately.

Case study: Enablement/empowerment in action

Business coach, Bob Bryant and Pat Wellington worked with a leading pharmaceutical organisation on a cultural change programme over the period of a year, initially working with senior and middle management.

Stage 1 *Management commitment*

The company created a project leadership team. It was composed of a director, and managers representing all the functions involved in the business: engineering, packing/dispatch, quality assurance, finance, personnel and training.

The purpose of the project team was to show management's total support and involvement for the changes that were to be implemented, and to develop the company's vision. The project team initially reviewed a range of options, and decided to develop their vision based on the concept of HOT (Honest, Open and Trusting) relationships. They then explored the way forward by creating a self-managed pilot team, and selected the packing and dispatch department to develop the new culture.

Stage 2 *The research phase*

A *cultural needs analysis* of the existing culture within the packaging and dispatch department was undertaken. Bob and Pat interviewed a cross-section of staff who had volunteered to be involved in the project. There were no management restrictions on the type of questions and information that could be collected.

A report was then prepared which defined the culture within the department from the viewpoint of the shop floor staff and from the management perspective. This enabled the gap between the existing environment and the vision of a company with HOT relationships to be defined. From Bob's and Pat's perspective this identified the size of the hill we were going to climb in partnership with our client.

A series of meetings was organised so that the volunteers, middle and senior management, could be introduced to the approach required to move to a self-managed team, the journey involved and the pitfalls that might need to be overcome. The management team then produced a document defining the principles of HOT relationships that they intended to develop in the packaging and dispatch department. A series of meetings was organised for the group of staff who had volunteered to be in the project. They were introduced to the philosophy so that they could make a reasoned choice as to whether they wished to continue to be involved in the change programme.

Stage 3 *Providing the pilot team with the hard and soft skills for the new culture*

The company made the decision to assume a no-knowledge base for everyone involved in the project from the packing and dispatch department, and we ran a series of workshops to underpin the culture that was to be introduced. This included how to work together as a team, liaising with internal customers, problem-solving and communication skills. These sessions brought together managers, supervisors, engineers and operatives.

Stage 4 *Letting the team loose*

The management then created the first two teams and let them loose, figuratively speaking, to go forward and learn by their mistakes. At all times senior management and Bob and Pat were in the background if the going got really tough. As is always the case when you trust people, they exceed your expectations. Because the teams set their own objectives they were not limited to the expectations of the senior management. For example, a specific team's objective was zero machine downtime. Whereas an experienced manager may have gone for 5 or 10 per cent improvement in this performance criterion, the team was also looking for 50 per cent improvement in productivity. All the targets the teams set themselves were stretching the boundaries.

The benefits to the organisation from introducing a new culture:

- Productivity improvement which averaged 32 per cent after six months, but had with some products gone as high as 200 per cent.
- Removal of the 'them and us' relationship between managers, engineers and operatives.
- Shop floor pressure on management to introduce changes that at first sight appeared to have no production benefit. Examples of this type of change were installing a telephone line for outside calls; computer terminals for operatives to examine management performance data; overtime being identified and approved by the team; a meeting room for the team to use, leaving the production line as and when they chose; the team to receive production requirements as soon as the production controller received them from sales engineers; and operatives being able to discuss issues with suppliers without checking management approval.
- Staff took the initiative to do things differently. Operatives used their relationship with warehouse staff to find out what stock was on hand. This enabled them to prepare for sudden requests from customers which could not be met by stock.
- Gossip between team members in the restaurant became positive.
- Shared training experience increased the rate of learning new skills.
- Team members had higher expectations in terms of team performance than the managers.

The success of the project, which led to the organisation rolling out the concept to more teams, was because managers took the step of trusting their staff. The staff responded to the trust with enthusiasm and commitment as can be seen by the following quotes from the team's first quarterly report:

'The team was full of enthusiasm and eager to get to grips with the two new lines. Immediately we hit our first hurdle; we had only one line ready to run. This was the first taste of "It's our problem, it's up to us to sort it out." So we did.

One week later the second machine was running with tooling made in-house by one of the team.'

'In the following weeks many hurdles have presented themselves and as a team we have overcome them.'

'There have been major benefits in areas, breaking down barriers by dealing with people who can help us resolve our problems.'

'The enthusiasm is still with us and the sky is the limit. The encouragement and support given from all areas will help us achieve our goals.'

A change in working relationships is never a simple rite of passage. Working with teams of all kinds in different industry sectors, we have found that once a pilot initiative starts to work and become successful, there is a contagious energy created within the organisation, and other teams embrace the concept of change far more willingly.

10.6   Chapter summary

- To enable/empower means to trust a team member or a team to make the right decisions, giving them freedom to do so and ensuring they accept accountability, but . . .
- . . . assuring them it is all right to make mistakes as long as they learn from them and modify future behaviour accordingly.
- Legitimise considered risk-taking and push down decision-making authority to the lowest level that risks are considered.
- Establish clear lines of support, and manage by exception.
- When delegating a task or project make sure staff are used effectively and identified processes occur.
- Delegation always involves a degree of risk – ultimately you are accountable for the activities in your team/department, even if a team member has made the error. *The buck stops with you!*

Chapter 11

Problem-solving activities/ quality improvement

In our fast-changing world, finding (the time for) solutions can be challenging. Problem solving requires a 'solutions' thinking attitude. As an individual, facts and knowledge can only go so far. Tough problem solving and quality improvement requires the ability to define the true problem, analyse the possible causes, create options, select the most feasible option and then implement it. This chapter will help you enhance your skills to find sustainable solutions and learn new systematic ways to approach problem solving to reach win–win decisions by increasing your awareness of problem-solving techniques and problem-solving tools.

11.1   Introduction

Problems have to be solved, crises have to be survived and opportunities have to be looked for. In each of these three situations there are special thinking needs. It is not good enough to regard 'problem solving' as a general approach which covers all three. An opportunity is a bonus which you first have to spot and then do. A crisis requires solution design rather than solution finding. Even within problems there are different types of problem, each with its own needs.

Most books on problem solving do not distinguish between a classic problem and a crisis. With a problem you have to get back where you know you should be. Thing may be going smoothly. There's a deflection that must be overcome. It's a simple matter of overcoming the interference and getting back on course. With a crisis we have to appreciate that there is actually a baseline shift and, as a result, a more fundamental change is required. A crisis is more serious than just a deflection. If nothing is done to reverse the trend there may be a disaster. Think of the 2008 credit crunch. Things are not going to be restored to what they were. Big problems are often called crises but they should just be called big problems.

'A pessimist sees difficulties in every opportunity; an optimist sees the opportunity in every difficulty.'

Winston Churchill

11.2   Problems are normal!

When you have a problem, do you go for a solution or a cause? Most people go for a (quick) solution and forget the cause. This is a mistake and is one reason why so many change programmes don't work. For example, how much of your time each week is spent reacting to problems, problems that recur regularly? Staff in many of the organisations we work with complain that they spend 60 per cent of their time handling such issues. This tells us that some manager some place is not doing his or her job of controlling change.

This chapter will set out to clarify a systematic approach to tackling problems. Flashes of inspiration are great but a more robust approach is required if you want to move from a victim to master of problem solving.

Problems, like change, are normal. The trick is to ensure that *reacting to* problems does not overtake you and your team. Finding solutions entails thoughtful planning and sensitive implementation and, above all, consultation with and involvement of the people affected by the problem.

What many team leaders and organisations miss is that when they go about solving problems they don't get to the root cause; they only treat symptoms. It is vital to get to the underlying reason why they have occurred to reach a successful outcome.

Problem solving is a better thing to do or a better way to do it. It increases an organisation's ability to achieve its goals. True problem solving is visible to others and must offer a lasting impact. Solutions have to be converted from an idea into action so that they work and can be sustained.

Using creativity and innovation to problem solve is very relevant to strategic management. It is the ability to solve problems that determines much of what an organisation is able to do.

Problem-solving capability provides the organisation with the bricks from which a strategy can be built. Creative thinking, when applied to a strategic situation, can provide the architecture that enables the bricks to be built into a bastion of strong competitive advantage.

Dr Kenichi Ohmae coined the phrase 'strategic thinking' and suggested that many organisations rely too much on analysis and not enough on creative thinking and creative problem solving.[1] Strategic thinking builds competitive positions and this requires creativity and willingness to innovate. It is a long way from the situation often found today where strategy is seen to be duplicating others in the industry rather than building a position based on creative thinking after sound analysis.

Drucker always emphasises the strategic importance of innovative problem solving and sees ideas as part of the process of making the future.[2] 'Tomorrow always arrives.' Organisations that fail to solve their problems, i.e. react (rather than control change) will suddenly find that they have lost their way.

So what are some typical quality improvements or problem-solving activities you might find going on in an organisation?

- A car factory engineering department: looking to improve road holding, reduced air resistance, less engine wear, reduced fuel consumption.
- A production department: improvements in cleanliness, materials maintenance, machine efficiency, zero production defects, no breakdowns.
- A marketing department quantifying output quality: price, assortment, attractiveness and service.
- A distribution centre: looking to ensure delivery on time, no wrong components, the right quantities.

So, in your situation how should you move forward?

There are four stages that you need to go through during any quality improvement or problem-solving activity (Figure 11.1):

Stage 1 *Research phase*
Stage 2 *Set goals – the 'vital few'*
Stage 3 *Decide the means to select the preferred solution*
Stage 4 *Monitor, implement and review*

Let us now examine each of these stages in more detail.

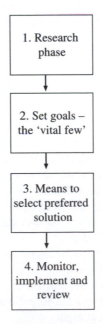

Figure 11.1   The four stages of problem solving activities

11.2.1   Stage 1: Research phase

This first phase of activity involves researching all your existing sources of information to identify potential projects. You could use a variety of techniques, for example:

From the sales and marketing arena:

- market research
- opinion/perception surveys
- product-user groups
- customer meetings
- Six Satisfaction Elements

From process monitoring:

- process mapping
- flow charts
- run charts
- Pareto diagram
- fishbone diagram

As engineers, you are probably very familiar with the process-monitoring approach, and likewise with the four top bullet points under the sales and marketing heading. Using elements from each to solve problems is not unknown.

The Six Satisfaction Elements is an approach that Pat has used successfully in organisations over the years. You might not be so familiar with this instrument, so we are going to clarify this approach in more detail.

Six Satisfaction Elements

It may be helpful to demonstrate the value of this approach by looking at the following case study. Pat worked on a customer service initiative with a prestigious private hospital.

First of all, Pat defined the brief with the CEO. She initially undertook research through the use of a cultural needs analysis and submitted a report with her findings. A series of workshops was delivered. One of these consisted of the top team and heads of departments looking to analyse what was of real importance to the hospital's customers (patients, specialists and consultants who used the hospital facilities). The Six Satisfaction Elements formed the core of this exercise.

A wide variety of influences affect customer satisfaction. These constitute the six elements as outlined in Figure 11.2.

In reality, although a customer may be happy with the product or service offered, if even one other element does not live up to expectations the customer will ultimately not be satisfied. In the hospital's case, while the in-patient experience was excellent, the car park was poorly lit and this led to patients and visitors complaints due to feelings of insecurity, especially at night.

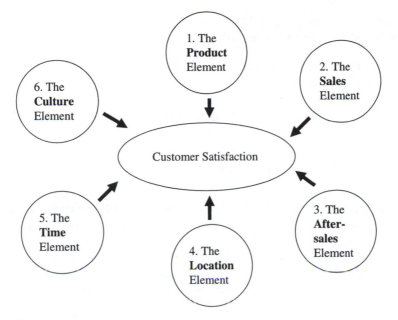

Figure 11.2   The Six Satisfaction Elements

For more information on these Six Satisfaction Elements, refer to Pat Wellington's book, *Kaizen Strategies for Customer Care.*[3]

11.2.2   Stage 2: Set goals – the 'vital few'

Once you have identified the improvements to make or the problems to solve, you must identify the 'vital few'. The vital few is based on the 80/20 principle. This principle was named after an Italian, Vilfredo Pareto, who established that 80 per cent of the wealth of any nation is invariably owned by 20 per cent of the population.[4] So, transferring this to the business world, you will probably find that most of your issues will come from a small number of causes. You might also notice that most of your complaints come from a small number of customers, not the majority.

You must set goals before you take action. The purpose of goals is to keep you on track, motivated, and focused on the actions you need to take. These goals need to be concrete (what exactly and by when exactly) or SMART or as has been more recently used, SMARTER.

SMART–SMARTER

Major term	Minor term
Specific:	giving a clear description of what needs to be resolved
Measurable:	in terms of quality, quantity and cost
Achievable:	including gaining the agreement of your manager and team

Relevant: specifying the business need to be satisfied
Time Bound: in terms of completion date

Ethical or Exciting or Evaluated
Reviewed or Rewarded

Exercise for identifying a SMART approach: do you really understand?

Here is a message from a team leader to his/her team. What is the problem with this and why do you think team members might find it difficult to understand? In the second box you will find the answers:

> Management has decided to outsource all the organisation's service activities, such as mail rooms, training, transportation, etc. to reduce costs by 7 per cent this financial year.
>
> The reason for this is a move to a concentration on core activities only, so that the organisation can manage its operations more effectively and satisfy shareholders by increasing profitability and overall cost effectiveness by 5 per cent.
>
> The benefits will be more specific delivery and service targets from these suppliers and will allow all personnel to concentrate on delivering their objectives.

Why was this communication not SMART?

> Management has decided to outsource **all [what does 'all' mean? Is there a list?]** the organisation's service activities, such as mail rooms, training, transportation, etc. to reduce costs by 7 per cent this financial year **[when does management expect these changes to start?]**
>
> The reason for this is a move to a concentration on core activities **[what core activities?]** only, so that the organisation can manage its operations more effectively and satisfy shareholders **[what does effectively and satisfy shareholders mean?]** by increasing profitability and overall cost effectiveness **[does management expect manpower to be impacted? Can we impact it?]** by 5 per cent.
>
> The benefits will be more specific delivery and service targets **[what are these exactly?]** from these suppliers and will allow all personnel to concentrate on delivering their objectives. **[How will objectives change?)]**

So to see SMART in action let's return to the hospital case example. The booking department clarified the areas that they wanted to improve and used SMART objectives to take things forward. Here are three examples.

Case study: How can SMART objectives be used?

Patient goal	Objectives	Lead person	Due date	Evidence	Spring progress review
Ensure we deliver the best care standard with resources available	Ensure there are no more than three justified, written complaints about the booking department in one year	P. Taylor	03/2008	Number of written complaints	
Consultant/ GP goal	**Objectives**	**Lead person**	**Due date**	**Evidence**	**Spring progress review**
Ensure we deliver the highest possible standard of service to consultants and GPs	1. Compile a directory of 'length of time' for standard operations, noting variations between consultants to assist in judging admission times and duration of theatre slots 2. Create a directory of consultant signatures to ease identification of outpatients' booking cards	P. Taylor with C. Hogarth	12/2008	Directory	
Building and equipment goal	**Objectives**	**Lead person**	**Due date**	**Evidence**	**Spring progress review**
Maintain facilities to the highest standards to ensure health and safety	1. Redecorate waiting rooms and reception areas. 2. Fix faulty drainpipes/ guttering	C. Jones	01/2009		

Since this booking department was in a hospital they used the term 'evidence' for monitoring the outcome. You could use the term 'achieved objectives?' in your matrix.

11.2.3   Stage 3: Means to select preferred solution

So, fine, you have clarified what the problem is, so now you need to think of a variety of approaches you might adopt to take things forward.

Practical problem-solving techniques

What are typical and practical techniques used to define or *analyse* problems in the earliest stages of problem solving? Here is a list of well-known and useful processes that can be used.

Brainstorming
Have a recorder list all the possible ideas from the group as quickly as possible without any evaluation of ideas. Gather as many ideas as you can as fast as you can. This list of ideas is then shortened and a final solution can be developed from the best items.

Brainwriting
Each person records an idea or solution to the problem on a piece of paper and adds it to a pile. Everyone then takes a different paper from the pile and adds an idea related to the one already on that page. They write down the first thing that comes to mind. These ideas can then be compiled and discussed to develop a final solution. (This is often successful in a quieter group, when it is difficult to get everyone talking.)

Nominal group technique
Each person shares their ideas. A recorder lists all the different ideas. Everyone then ranks their preferences individually from the whole list from one up to ten. These scores are added together and a group score is given. This gives priorities on a group basis. (If number one is used to rate an individual's best choice, then the list item with the smallest group score is the most desired.)

Force-field analysis
List forces 'pushing for changing the situation' and 'forces pushing against changing the situation'. Outline the strategies to minimize the strength of the forces and develop an action plan to accomplish the solution.

Criteria matrix
Develop a set of standards that each alternative is judged on. Some examples are: costs, risks involved, timeliness, convenience, satisfaction. The matrix looks like a table with the alternatives down the left side and the criteria across the top. Each alternative is ranked by the criteria (1–5, yes or no, etc.) and the scores are added up at the bottom.

Dotmocracy
Provide each person with the same number of dot stickers, pennies or tokens. Vote individually on the list of brainstormed alternatives. If someone feels strongly about one item they are welcome to put all their dots on that item. The alternative with the greatest number of dots is the decision preference.

Bubble-up/bubble-down
Used for ranking statements and ideas or for putting ideas into a orderly sequence. Read the first two statements, decide which statement is least preferred and should be eliminated. Now compare the statement left to the next one on the list and continue to do this until you have reached the last item on your list and the group is satisfied. (For example, when purchasing a house it is much easier to compare the house you're viewing with the last house you viewed. After evaluating how this one rates, you eliminate one of the two. The preference of the two is the only house used in future comparisons.)

Circle chart
Divide the paper into four quadrants and label each quadrant one of four titles: What is wrong? (in theory), What might be done? (in theory), What is wrong? (in the real world) and What might be done? (in the real world). This will initiate a brainstorming session to help develop a solution. In reviewing all the expectations and priorities a solution can be chosen. But a poor decision is better than no decision. Set up an action plan to achieve your solution.

Lateral thinking
There is a variety of lateral thinking or problem-solving instruments you and your team can use to develop creative thinking. One is de Bono's *Six Thinking Hats* approach.[5]

So let's look at lateral thinking in more detail. Lateral thinking involves an element of provocation. With provocation we say something in order to jerk our minds out of the usual pattern. In normal thinking we use judgement but in provocative thinking we use 'movement' instead.

Lateral thinking is a way of solving problems by apparently illogical methods. De Bono defines lateral thinking in illustrative form as: 'you cannot dig a hole in a different place by digging the same hole deeper'. This is like a politician saying there is 'no point doing the same thing over and over and expecting a different result'. Trying harder or putting more effort into doing the same things may not be as useful as changing direction.

For example, in de Bono's *Six Thinking Hats* approach, many team members get bogged down at the data and information stage, and keep asking themselves 'what information is missing?' They forget to move to stages two and three which are all about using their intuition, 'following a hunch' or using their critical judgement. All too often we have observed teams getting bogged down by negative thinking at one stage and not using lateral thinking to move through all the necessary stages to make progress.

Lateral thinking is different from and much more than problem solving. As de Bono points out, problem solving implies there is a problem to respond to and that it can be resolved. That eliminates situations where there is no problem or a problem exists that cannot be resolved. In lateral thinking it's usually good to think about situations where there are no problems so that we can make them even better. Sometimes, too, we may need to solve problems not by removing the cause but by designing the way forward even if the cause remains in place.

Example: Injured boy problem

A father and son are passengers on a train that crashes. The father is only slightly hurt but is pinned in the wreckage. His son is unconscious and is taken to hospital for examination. The doctor, upon seeing the boy, is visibly shaken and says, 'I can't examine him or operate. The boy is my son'.

Problem – What is the relationship between the doctor and the boy?

Lateral thinking is a method of thinking concerned with changing concepts and perception. Lateral thinking is about reasoning that's not immediately obvious and about ideas that may not be obtainable by using traditional step-by-step logic or traditional thinking. Traditional thinking (as in the *Injured Boy Problem* above) can lead to standard solutions that may not work in our fast-changing world. For lateral thinking to work we must suspend our traditional thinking methods.

The *Injured Boy Problem* is an example of an instant perception blocking the mind's ability to explore alternatives. In this case the instant perception is that most people imagine a surgeon as a male. If you switch your perception to allow for a female surgeon then the answer is suddenly obvious; the surgeon is the boy's mother. Lateral thinking is the method of switching perceptions to allow the alternate viewpoint.

According to a 2004 study by Bill Breen ('The six myths of creativity'), creativity does not come from special types of people.[6] Creative thinking can apply to everyone. The ability to think in new ways and the ability to push through creative 'dry' spells is crucial. Intrinsic motivation is crucial. You must want to do it. What stops us? The usual excuse that we don't have enough time.

To change anything you need thinking time. When is the last time you really thought about what you want to achieve at work or at home? Since we are all so task driven we forget to stop and think. The report also highlights that monetary rewards can have a negative affect on creativity by making people risk averse. Most creativity gurus say that stretching people's skill levels can be reward enough. When is the last time you were pushed to realise something you didn't think possible? Did it make you feel proud of yourself?

Many people are given such tight deadlines today that creativity suffers due to the lack of 'incubation thinking time'. As observed in Chapter 5, this report also says that internal competition fosters creativity. The most confident teams share and discuss information. However, in the finance and high-tech industries there is a belief that internal competition fosters innovation. The report found that the exact opposite was true. When people compete for recognition, creativity can stall as people stop sharing ideas and information. This leads nowhere from a creative perspective as no one individual has all the answers.

11.2.4   Stage 4: Monitor, implement and review

It is important to have commitment to your plans before any action is put into motion. Monitor the results of your decisions and their impact on others regularly. You may need to make minor changes as you go.

You must also consider the need to monitor your environment. You need to know *early* if any risks to implementation show signs of occurring. Your monitoring programme *must* always include regular communication with all stakeholders and with the leaders of any projects you are depending on. You will want to let these people know either that everything is on track or, if risks are emerging, that you are aware of them and that you have an appropriate contingency plan and understand what implications there might be for others.

To conclude, what are the practical questions to ask during the review stage? Always ask your questions positively. Never threaten or blame! What's done is done. As was said in Chapter 3, the responsibility of a team leader is to control change, i.e. your responsibility is to ensure that mistakes do not recur and that problems are solved quickly and decisively. How can this be done? By constantly asking questions to find solutions and capture learning. A five-minute chat can make all the difference.

- What's going well or what did you set out to achieve?
- What actually happened?
- Why did that happen?
- What should we do differently next time?
- What action should we take?

When things don't go to plan

Many engineering managers manage people who manage projects. Increasing our understanding of how best to mentor project managers can help engineering managers become more effective. Project success has been described as similar to a three-legged stool with one leg representing project managers, one leg representing line managers and one leg representing top management. Project managers and line managers are considered equals in the organisation and

> projects often require working across boundaries that are difficult to navigate. The third leg of the stool, top management, is critical and necessary for project success. If the third leg is dysfunctional, then project success is unlikely.[7]

We've heard it many times before, and it's still true: if something can go wrong, it will. You must learn to figure out ahead of time where your problem-solving effort is vulnerable and develop appropriate contingency plans. Start on this as soon as you begin the problem-solving effort, making it a normal part of defining a problem.

Vulnerabilities (risks) are all the things that can prevent your problem-solving project from succeeding. Typical vulnerabilities include changing priorities, inadequate resources (people, money, time), lack of senior management sponsorship, staff turnover, key players unable or unwilling to participate, other projects not getting completed on time, economic change, etc. Of course, the list will be different for each problem, and the probability that any particular vulnerability will occur varies as well. The key is to identify them, and assess each one for probability of occurrence and impact on the project if it should occur.

Develop contingency plans immediately for any that have both a high probability of occurring and a high impact. You may also want to develop contingency plans for low-probability/high-impact issues. Low-impact issues, especially if the probability is low, are probably not worth significant contingency planning.

And finally ... Six Sigma

In the last few years Six Sigma has been adopted by many organisations with great success. The approach was first used in 1986 by Motorola USA as a statistically based methodology to reduce variation in electronic manufacturing processes. It has subsequently been used by organisations such as GE, where in 1998 they claimed that Six Sigma had generated over three-quarters of a billion dollars of cost savings.

Six Sigma is commonly recognised as a tool for understanding statistical variation within products and processes. The aim is to reduce this variation to produce more consistent functionality or process output – leading to better processes, more reliable products, lower costs and ultimately happier customers.

At the heart of this methodology for process improvement is the acronym DMAIC:

D   Define opportunity
M   Measure performance
A   Analyse opportunity
I   Improve performance
C   Control performance

It is based on a rigorous and disciplined methodology that uses data and statistical analysis to measure and improve operational performance by identifying 'defects' in manufacturing and service-related processes. However, it is more than a statistical tool. Everything we explore in this book with regard to leadership style, communication and behavioural approaches is critical during the implementation of Six Sigma.

Adrian Tan, a Black Belt (certified expert) colleague of Pat's from Singapore, shared his thoughts on why and how projects can falter:

Training Lean Six Sigma Green Belts to lead improvement projects

I have been providing training for Lean Six Sigma (LSS) Green Belts for some time already, involving participants from a variety of professional backgrounds. While developing staff and providing advice for Lean Six Sigma projects, I have noticed that a successful Lean Six Sigma deployment is truly dependent on improvement projects which are led by good project leaders (Green Belts). The project leader plays an essential role in mastering the right problem-solving strategy to lead the team in achieving the desired project objectives. Projects tend to fail to deliver not because the leader is deficient in knowledge of a certain subject matter, but due to his or her lack of mastery of team dynamics and poor project management skills.

Once, I was coaching a group of Green Belt wannabes in leading their improvement projects with a team. It was through this live coaching of projects that I experienced a spectrum of project leadership styles with varying degrees of successes.

On one side of the spectrum were project teams led by leaders tending towards the 'I know it all' or 'I know better than you all do' attitude. What happened in their case was a team that wasn't a team. Team members were nonchalant towards the project objectives and, needless to say, the project leadership. They were not empowered to air their ideas and even if they were, they faced rejections from the project leader. It became natural that the team slumped into what I term a perpetual 'stunned silence'. No-one wanted to contribute in any discussions as they were sure that their views were never going to be valued or at the very least evaluated. The result of such a team phenomenon was what I would call 'team degeneration'. Ideas were there, but only from one person – the project leader – but which had limited progress in implementation and resistance from stakeholders.

On the other end of the spectrum was what I call a 'free-for-all' team. In this project team meetings and discussions were informal and friendly, but tended to deliver no conclusions or progress. There were bountiful ideas provided by the team, but the team leadership was exceptionally weak in aligning ideas to the objectives of the project. Ideas tended to veer off course and seldom got back onto the main track. The result was a problem that never got solved. There was never a start and an end.

Encountering these team phenomena, I tried to coach the teams and their project leaders to use the Lean Six Sigma DMAIC roadmap to lead and guide their improvement projects. It proved successful. They managed to use the roadmap along with the various decision-making tools, from project chartering and scoping to control and closures, to drive their improvement projects to implementation with desirable results. The good thing about using this roadmap is that the project leaders have a visible path to guide the teams in empowered decision making. By using the roadmap, it was observed that project leaders were better able to lead their teams in problem definitions right down to the root cause, solutions generation, selection and implementation. They were also able to institutionalise project milestones with the team, thereby completing their projects on time, and within scope and budget. The most interesting observation on the project teams using the DMAIC roadmap was that they became more open-minded and motivated, and were more focused in their contributions towards the project objectives. This is truly one of the powers behind the Lean Six Sigma DMAIC roadmap in being able to nurture high-performance self-directed teams.

Adrian Tan
Managing Partner, One.Strategist LLP

11.3   Chapter summary

- We recommend a four-stage problem-solving approach, always ensuring that there is a monitoring process in place.
- Six Satisfaction Elements is an optional approach to use in problem solving and quality improvement activities.
- Select the 'vital few' and use SMART goals to achieve your objectives.
- Encourage creative thinking and keep a close eye on possible risks during implementation.

Part 4

Communication strategy

Chapter 12

Effective communication

There are a variety of ways in which you communicate with your team, with others within your organisation and externally with clients and stakeholders. Communication can take the form of written words, conversations on the telephone, via new media such as email and videoconferencing or face-to-face contact. This chapter will explore all these forms of communication plus some of the barriers that can inhibit good communication taking place.

12.1 Introduction

Just think of all the different methods you can now use for communicating or gathering information:

- project meetings/report writing and planning
- telephone contact/face-to-face contact
- Skype/instant messaging/WebEx
- intranet and email
- Wiki pages
- knowledge collection portals
- real-time conferencing

Collecting information to undertake a task can be comparatively straightforward; the tricky bit kicks in when you need to communicate.

We are all aware that communication is happening all around us. There are communication managers or departments in organisations dedicated to communication/PR. Communication skills are a key competency looked for during the recruitment process. We are constantly being told we need to be 'good communicators'. So this chapter is dedicate to the topic.

The word communication is derived from the Latin *communis* meaning common, shared. Until you have shared information with someone you haven't communicated it. In addition, the person you are communicating with has to interpret the information, and there needs to be a shared understanding between both parties.

As a starting point it is important to realise that there are several barriers which can reduce the effectiveness of communication.

'The single biggest problem with communication is the illusion that it has taken place'.

George Bernard Shaw

12.2   Barriers to communication

Barriers to communication include:

- **Differences in perception** – Our way of viewing the world depends on our backgrounds. People of different ages, nationalities, religions, cultures, education, status, sex and personality will perceive things differently.
- **Language** – A range of people will talk to you differently – some simply, some by using jargon, some by using complex sentences.
- **Assumptions** – Because of preconceived ideas, we may see or hear what we were expecting to see or hear instead of what was actually done or said. Jumping to conclusions is a very common barrier.
- **Stereotyping** – Because we learn from experiences (good and bad) there is a danger of adopting set attitudes, e.g. 'all teachers are the same. They think they know everything!'
- **Lack of knowledge** – If the person trying to communicate is not really sure of him/herself or the receiver does not have the pertinent background information to make an informed decision, communication can be challenging.
- **Lack of interest** – If the recipient is not interested, the communicator will have to work hard to make his/her message appealing.
- **Problems with self-expression** – Some people find it difficult to express what they really mean because of a limited vocabulary or a lack of confidence.
- **Emotions** – Emotion can be a good communication tool and also a barrier. If the emotion is too powerful it could distort the message you are trying to communicate. There may also be a hidden agenda.
- **Personality** – A clash of personalities is a common barrier to effective communication. You may not be able to change the personality of the recipient but you should be able to control your own behaviour. A lack of mutual trust, respect or confidence can have a detrimental effect on communication.
- **Attitude/environment** – If people are in an environment which does not encourage free exchange of ideas, communication can be inhibited.
- **Feedback** – If it is not possible to get instant feedback it is difficult to check whether or not our communication was effective.

12.3   Breaking down communication barriers

So, bearing all of this in mind, the objectives of your communication should be to understand:

- What type of person is s/he in terms of personality, education, etc.
- How will he/she react to the communication?
- How much knowledge does s/he have about the topic?
- How much time will s/he have?

Where will the communication take place?

- Will the recipient be close enough to any relevant information or in an isolated situation?
- Will you be able to deal with any queries easily?

When must the communication be completed?

- Will the recipient be too busy?
- Is the deadline realistic?

What is the subject and purpose of the communication?

- What exactly needs to be said?
- What does the recipient need to know?
- What can be omitted?
- What information must be included so that the objective(s) of the communication are achieved?

How should the communication be effected?

- What method of communication should be used?
- How should the points be organised to ensure that a logical structure is maintained?
- How can the recipient's interest be maintained?
- How can the objectives be achieved?
- How can it be ascertained that the communication has been understood? Is the method of feedback suitable?

With regard to people making assumptions: differentiate between fact and inference. Inferences that get turned into facts during their transmission can have serious consequences. They often cause false rumours. An inference represents only a degree of probability. It goes beyond what you are actually observing and draws conclusions that are not necessarily true, e.g. if you see a doctor's car parked outside your neighbour's house and you say, 'Somebody's sick over at the Smiths', you are stating an assumption as if it were a fact. You would be more accurate if you said, 'There's a doctor's car parked in front of the Smith's [the fact]. Maybe somebody is sick there [the assumption]'.

In addition, to clarify information:

- Keep the number of links in the communication chain as low as possible.
- Use more than one medium for important messages.
- Limit the number of items in a message.
- Use illustrations and sketches to reinforce messages.
- Itemise the points and put them in a logical order.
- Highlight the most important points.
- Use associations that will help the recipient to understand the message.

So, having looked at communication barriers and steps to take to overcome them, let's move on to the three main categories of communication: putting it in writing, communication on the telephone and face-to-face contact.

12.4   Putting it in writing

Communicating in writing obviously has an advantage over face-to-face contact, since you have things in black and white. This can work in two ways – to your advantage or to your detriment. The advantage is that you have proof of your thoughts and proposals. The disadvantage is that the written word is permanent and can be misread or misunderstood.

So how can you make things clear, so they are interpreted in the way that you intended?

Looking at written communication in general, how readable are your proposals, reports, internal publications, newsletters, brochures or handouts? If you think of reading a legal document, say a mortgage agreement, how riveted are you with the copy? Do you understand it? How about the latest thriller by your favourite author? Why do you feel such a change in state when you think of these two written documents?

The answer lies in readability.

The 'Fog Index' gives a measure of readability based upon the number of years' reading experience a person must have to understand the content.[1] It calculates using the average number of words per sentence and the number of 'long' words (three or more syllables).

To put this into perspective here are some comparisons:

Mr Men books	3–5 years
Modern action fiction	10–12 years
The Times editorial	15–20 years
Legal documents	25+ years

So what do you need to bear in mind when creating written communication?

- Keep sentences short, 15–20 words maximum. This makes the copy more readable and digestible.
- Be wary of the jargon you use. A series of initial such as CRM/MDU might make sense to someone within your organisation, but externally to a client they might be meaningless.
- Use 'spoken' writing. Think how you would express yourself if you were talking to someone, and write what you would say. Use everyday words.
- Avoid pomposity, and trite or over-used phrases, for example, 'it is my considered opinion ... ', 'it has come to my notice ... '.

For letter writing:

- Establish your purpose – am I writing merely to state facts, give additional information, or persuade?
- What do I want the reader to do? Are there action points that need to be highlighted?
- Put yourself in the recipient's shoes. Think what is important to them, and use 'you' as often as possible rather than 'I'.
- From a layout point of view, include the main points in your first (short) paragraph, go into further detail in the middle paragraph, then summarise, giving clear guidelines if actions are required or timing involved.

When writing memos, it is important to be direct; don't be pompous or fudge the issue. The recipient does not need to have a lengthy explanation as to why you might need to cut costs or relocate staff to a different part of the building. Get to the point in the first sentence. Don't beat around the bush and slip a request into a final paragraph as if you are embarrassed about the situation.

12.5   Communicating within your organisation: make it appropriate

For routine written communication within your organisation you need to bear in mind the following:

- More senior staff don't need to know 'nuts and bolts' detail. The information you supply should be strategic and 'big picture', such as the positioning of your organisation in the marketplace, opportunities and threats.
- Your peers should be given tactical information such as how much it will cost, what do you ask or expect from them, and how will it help your teams to work together more effectively.
- Team members want practical details, such as where they can gather information, who will be working together on a project and what budget will be allocated.

- Tailor your style of writing and make sure that what you are creating is what the recipient expects to receive. Sometimes you can make a mistake in what you produce if you don't ask the right questions at the time of the request for copy.

Pat learnt this the hard way. When she first headed up the management development unit of a London university, the Dean of the Business School asked her to create a business plan for the unit. She made the cardinal mistake of not asking how comprehensive the plan should be. She produced a sizeable document of thirty or forty pages which took a considerable time to create. As her template, she had taken a business plan that had been passed to her by the head of the international department, which had a substantial turnover in comparison to the business unit that she was running. It turned out that the Dean had expected a brief report of maximum five to ten pages!

Case study

You need to sense the style of communication that other colleagues wish to receive, and be particularly careful when you first join a company or when a new person takes over a team or department. The written word is powerful, and you can really upset the applecart if you decide to tackle a deep-seated problem with another business unit by writing a strident memo to the manager of that unit before you have even met him/her face to face.

This is an example of a breakdown in communication between two individuals that we discovered during one of the needs analysis assignments undertaken within an organisation:

It had taken the sales manager in this organisation several years to build a up good client base, but in the last six months since the new finance manager had come on board, several of his key clients had experienced problems with the accounts department. The problems were as follows:

- Inaccurate invoicing sent to a client's former address (even though the sales department had informed accounts of the new address).
- One of his clients had requested an invoice to be sent from his company prior to the end of their financial year so that they could pay for the services they had received within that year. One of the account managers in the sales department had checked with someone in the accounts team and had been assured that this had been actioned. The account manager then heard from the client that he had not received the invoice in the timeframe specified. She complained to the new finance manager and received an inappropriately aggressive response from him, saying that he was short staffed and no-one had the time to do it.
- The final straw for the sales manager was when he heard that one of his major accounts was threatening to withdraw their business from the

company because of the way they had been treated by a credit controller in the finance department. The credit controller had handled a late payment of an invoice in an over-aggressive manner and accused the client's accounts department of not paying when in fact the payment had been made in the previous month.

Now all of these errors were small in their own right, but the end result was the possible loss of a substantial amount of revenue for the organisation.

The big mistake that the sales manager made was to write a strident memo to the new finance manager giving little or no details of the actual problems that his department had experienced in the past few months. Instead, the tone of the memo accused the finance department of putting sales at risk, and accounts staff of being rude to clients.

This approach was not well received by the new finance manager, and a 'war of words' continued for quite some time, all in writing, without the fundamental problems being resolved.

So how could this situation have been resolved in a more satisfactory way?

Face-to-face contact should have been established between the two internal customers when the new finance manager came on board. It wasn't, and strident memos between the two parties did not get their relationship off to a good start.

The sales manager should have presented his case in a more factual way. Staff from an analytical finance background understand data, flow charts and revenue. It therefore would have been prudent for the sales manager not only to present the hard facts, but also the long-term financial implications of losing the named key business accounts.

Business units can often be under pressure and understaffed; year-end in the finance department will usually be a pressured time of year, with various business units asking for speedy responses for their clients. Not that this is an excuse for requests to be ignored, but if you as a team leader have established a good working rapport with a department you have a better chance of your request being heard, actioned and given priority.

There is always the dilemma of how much information to provide in writing in advance of a troubleshooting meeting. Of course facts need to be presented, but the question is when and how? All of the facts in this case should have been highlighted in advance. However, as no face-to-face contact had been established between the two parties, the finance manager could still have become defensive at the meeting from the word go.

An initial icebreaking meeting could have been set up by the sales manager, indicating that problems had occurred between the two departments, and that he would bring details to the meeting. Factual information could have then been presented face to face, where there was time and an opportunity to build a rapport between the two parties.

So, as a team leader what can you learn from this case?

- Face-to-face contact with key internal customers is vital, particularly when someone new joins the organisation.
- Be careful how strident you are in memos – it can come back to bite you!
- Understand the needs and requirements of your internal clients, by building a rapport with them and asking them about the flow of work in their unit – when are their pressure times, and how much lead time they require in particular situations.

12.6   The use of email and videoconferencing

Email has revolutionised the way we communicate with each other. It is quick, cheap, gets over the barriers created by time zones ... and can be misused.

George Bernard Shaw said, 'the more sophisticated our technology, the less we communicate'. How true this has become. In many organisations human contact seems to have be lost. All communication is by email – even people sending emails to a person sitting next to them.

As a team leader it is important for you to establish with your team the balance between using email and having meetings, formal or informal. Establish as a rule of thumb guidelines as to who should be copied into a mail message. So many organisations are blighted by email overload, and it becomes impossible for staff to see the wood for the trees. This can often be a cunning ploy to camouflage bad news: the 100-page report sent as an attachment telling staff on page 98 that various departmental budgets are going to be cut ...

Email can be good for passing information on speedily, but it is not so good for dealing with emotionally charged situations, or clarifying a deep understanding of a situation. Here are couple of guidelines with regard to creating email copy:

- Don't be tempted to use text writing; not everyone can understand it.
- Be careful how you use humour; it can be misinterpreted by the recipient.

How about videoconferencing? What are the strengths and weaknesses of using this form of communication? The obvious strength is that you can get a group of colleagues together on different continents and they can conduct a meeting. It is a step up from telephone conference calls.

TelePresence, developed by Cisco Systems, creates face-to-face meeting experiences over the network, letting people interact and collaborate. *TelePresence* refers to a set of technologies which allow a person to feel as if they were present, to give the appearance that they were present or to have an effect, at a location other than their true location. In essence, you can get staff around a number of tables around the world and they can talk to each other as if they were sitting face to face in the same room.

Many videoconferencing facilities are not as sophisticated as this product. The problem will be that you cannot look into the remote person's eyes and tell where they are looking, so you cannot tell if they are paying attention. With some videoconferencing facilities there can be at least a one-second transmission delay which takes away cues and makes feedback out of kilter.

12.7   Communication on the telephone

If you do not have the benefit of seeing the person you communicating with, there are a few considerations you need to bear in mind. You might get away with a hurried or slurred manner of speaking face to face, but on the phone it can really be confusing to the recipient. It is important therefore to be aware of certain aspects of voice-craft when using the telephone.

Pace

Talk at a slower pace than normal. The telephone exaggerates the rate of speech. A good guide is an average of 150 words/minute. Remember that your listener could be hearing the voice of a stranger and he or she needs time to accustom themselves to you. If you speak too rapidly it is harder for the listener to understand, and misunderstanding can lead to mistrust. Similarly, if you speak too slowly the listener can become impatient and irritated.

Pitch

The pitch of your voice will vary with your state of mind. If you are excited or impatient the pitch usually rises. Because the telephone exaggerates pitch, it is advisable to keep to deeper tones; it gives you more authority.

Emphasis and inflection

Interest and understanding can be enhanced by emphasising certain words. Inflection rises if you are expressing doubt or asking a question. When the meaning of a sentence is complete, inflection drops. Most people subconsciously listen to a speaker's inflection as this often signals they are coming to the end of a sentence, so it gives them the opportunity to interject.

12.8   Face-to-face contact

Why is it that you can walk into a room, and instantly be able to talk with ease to a particular person even if they are a stranger to you? It is because you feel you have built a rapport with that person. Building rapport is essentially a pattern-matching process where both people are on the same 'wavelength'. Most rapport building can be without words – it is how a person projects himself through body language, facial expression and tone of voice. If there is a mismatch between words and body language then you instantly believe what body language is telling you.

When you meet another person, both of you will send and receive lots of 'signals' or 'messages' about the way you feel and what you think of the other person(s). An assertive person will demonstrate positive body language, with open gestures which radiate a sense of confidence and readiness to listen to the other person. By contrast, an aggressive person will demonstrate negative body language and will use gestures and behaviour which 'put down' the other person. Likewise, a submissive person will demonstrate a diffident form of negative body language, which signals his/her feelings of low self-esteem and general lack of confidence.

	Assertive	Aggressive	Passive
Posture	Upright/straight	Leaning forward	Shrinking
Head	Firm not rigid	Chin jutting out	Head down
Eyes	Direct no staring: good and regular eye contact	Strongly focused staring, often piercing or glaring eye contact	Glancing away, little eye contact
Face	Expression fits the words	Set/firm	Smiling even when upset
Voice	Well modulated to fit content	Loud/emphatic	Hesitant/soft trailing off at ends of words/sentences
Arms/head	Relaxed/moving easily	Controlled/ extreme/sharp gestures/fingers pointing, jabbing	Aimless/still
Movement/ walking	Measured pace suitable to action	Slow and heavy or fast, deliberate, hard	Slow and hesitant or fast or jerky

When people meet us for the first time, 50 per cent of the impression they have of us comes from body language. If, for example, a person is standing quite defensively with their arms crossed, and their eyes either riveted to the ground or wandering around the room and not giving you eye contact at all, it is pretty hard to feel you can build a good rapport with that person. Going into a person's office to talk to them and finding that they don't stop what they are doing, possibly have their back to you pinning things on to a wall chart, etc. can also be off-putting, and make you feel you want to get out of their space as soon as possible.

So, what can you do to make a person feel at ease in your presence? As you can see from the matrix above, your body language should be as relaxed as possible, arms held loosely either side of your body, and making eye contact with the other person.

The tone of your voice can also make a difference. If the pitch of your voice is too high it can come across as strident and off-putting. If you lower the tone of your voice and speak in a softer tone you will create rapport more easily.

A good technique in conversation is to match the other person's communication style. Matching is not the same as mimicking or copying. It is getting in tune with the energy level of the other person's conversation. If the person is in a hurry and wants something to happen promptly and at speed, an energetic minimal response is all that is required. If someone is unhappy and wants you to comfort them, space and time to think for them is an important factor. We often do these things instinctively; however, there are other techniques that you can use in a conscious manner.

During a conversation, you can pick up different signals about people's preferred style. Are they visual, auditory or kinaesthetic individuals? What sort of wording might they respond to?

- If an individual is a visual person, using phrases such as 'I see', 'clear-cut' or 'beyond a shadow of a doubt' will help you communicate more effectively with them.
- Auditory individuals tune in well to expressions such as 'loud and clear', 'I hear you' or 'describe in detail', etc.
- How about kinaesthetic individuals – those people who are good with their hands, or maybe physically skilled using their whole body? They will relate well to terms such as 'come to grips with', 'pull some strings' and 'hold it!'

A further language consideration is based on Eric Berne's transactional analysis theory and model.[2] He said that each person is made up of three alter ego states: parent, child and adult.

The language of parent is based on absorbed information and conditioning from when we were children. Parents, teachers and older people use wording such as 'under no circumstances', 'never', 'poor thing', 'there there', etc. The language of child can manifest itself as baby talk, like 'I don't care', 'I want', I dunno'. Adult communication is based on our ability to think and determine actions for ourselves.

When we communicate we are doing so from one of these ego states, depending on our feelings at the time. There is no general rule as to the effectiveness of any ego state in any given situation. If we use too much of the parent mode of communication – particularly if it moves into a critical parent mode, then this can stifle and inhibit good communication.

Some people get results by dealing with others in the parent-to-child mode, or by temper tantrums in the child-to-parent mode. As a rule of thumb, if you are looking to build a rapport with another person as quickly as possible, using an adult-to-adult mode of contact is preferable.

So how does an adult mode of contact manifest itself?

- A lively facial expression, actively listening and offering an appropriate response to what the other person is saying.
- Using terminology such as 'why is that', 'how come', what, when, where, who, how, in order to get the other person talking, to identify if the other person is stating an opinion rather than a fact.
- Restating what the other person has said and identifying it as a restatement to check for mutual understanding.

So body language, tone, and use of wording can all make an impact on how you are perceived, and how well you build a rapport with another person.

12.9   The importance of listening

Warren Bennis said, 'For most of us, thinking that we have 'tuned in' to the other person, (we) are usually listening most intently to ourselves.'[3]

Most people are better at talking than listening. At school or university talking is actively promoted in the form of debate. As a student you are encouraged to take a view on a topic and vigorously defend your stance, convincing others of its worth, and attacking any view that is in conflict. The problem with this form of communication is that it is adversarial and sets up a boxing match between competing opinions. Arguing can stop you questioning your thoughts and discovering common ground with another person.

There is a saying, 'we hear with our ears and listen with our brains'. Listening is different from hearing. It involves trying to understand what the other person is saying. When listening you are interested not only in the words, but in the tone of voice, what is left unsaid, and the other person's body language.

Most people, however, don't listen – they just take turns to speak. We all tend to be more interested in expressing our own views and experiences than really listening to the other person's point of view. This is ironic as we all like to be listened to and understood. Stephen Covey rightly said that when we are understood we feel affirmed and validated. In his own words, 'Seek first to understand, and then be understood.'[4]

> 'The wise man is not the man who gives the right answers: he is the one who asks the right questions.'
>
> Claude Levi-Strauss, *The Raw and the Cooked*[5]

Listening is the most used skill (45 per cent of our time), and in our schooling the least taught (see Figure 12.1). Most people who are not actually talking are busy rehearsing what they are going to say next.

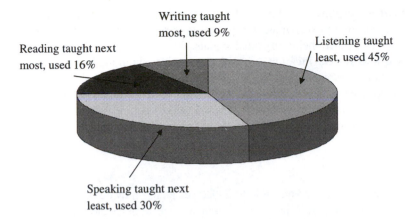

Figure 12.1 Use of listening and communication skills in schools

There are five listening levels, with I being the minimum and V the maximum:

I	Not hearing anything. Responding incorrectly and/or incompletely. Minimum involvement in the communication exchanges.
II	Listening but only hearing some things. Responding incompletely. Deeper meanings are not heard. Superficial involvement.
III	Hearing most things. Responding correctly to all that is heard. Adequate involvement in the communication exchanges. Known as *passive* listening.
IV	Hearing everything said and fully responding.
V	Hearing everything said and not said, i.e. *hearing between the lines*. Responding as a result of correct interpretation of verbal and non-verbal language. Wholly involved in the communication exchanges. Known as *active* listening.

So why are we so often bad listeners?

First, the speed at which we think. The average person talks at about 125 words/minute. The brain can think at speeds of up to 500 words/minute. We get distracted.

Second, there are the problems of outside distractions, of tiredness, discomfort, noise, movement, phone calls, interruptions, etc.

Third, we can all interpret information in different ways. Good listening is active rather than passive. So how can you let another person know you are listening? There are a number of ways in which you can do this.

- **Posture** – Face the other person squarely in an open posture, lean slightly towards the speaker and try to be relaxed and keep good eye contact.

- **Neutral responses** – When listening, use non-verbal or paralinguistic signals (mm, ah, ow) to encourage the speaker to carry on talking and to show you are listening. Watch their non-verbal signals.
- **Requests for clarification** – If you are not clear about the meaning ask for clarification or explanation.
- **Paraphrase** – Restate the message in your own words, to show that you have understood correctly and to let your colleague see your interest.
- **Positive response** – Summarise and give positive reactions before giving negative ones.

> 'We have two ears and one mouth so that we can listen twice as much as speak.'
> Epictetus (Greek philosopher associated with the Stoics, AD 55–c.135)

The fundamental capability you should have as a team leader is to ask the right questions, stimulate thought, and get others to think for themselves.

Pat was talking to a colleague of hers, Bob Bryant, who worked as a CEO in two NHS trust hospitals in the West Country. He said that in many organisations an 'interrupt' culture exists. This will often be a senior person interrupting a more junior member of staff, consciously or subconsciously, considering their point of view to be more important than the other person. Or, it could be that one person is too enthusiastic to put across their point of view. Bob said:

> It was not until I had left my role as a Chief Executive and became a business coach that I found out about my personal listening skills. They were terrible. My staff would be half way through telling me about a problem, and I would be interrupting with the solution and getting on with the next problem.
>
> Then Charles Smith from America introduced me to the Indian Talking Stick. This is a stick that Indians placed in the centre of the tent when they were having a tribal conference. The person who wanted to speak would pick up the stick and while they were talking no-one could interrupt, ask questions, make comments or disbelieve what was being said. The person holding the stick was tribal chief until he relinquished it.
>
> I tried the talking stick with a group of people I used to work with. The rules we applied were simple: no interruptions, no questions and no comments until after the meeting was over.
>
> Subsequently we analysed the experience. The people involved believed they had achieved four hours' worth of communication in a single hour. They felt that not having to worry about answering questions enabled them to concentrate on the message they were trying to get over. Knowing there were not going to be any clever remarks to put them down meant they did not

have to be ready to defend themselves. No interruptions meant that they actually said what they had to say.

As far as the listeners were concerned they really listened to the WHOLE statement rather than switching off while they thought of something to say.

12.10   Chapter summary

- Be aware of communication barriers and do your best to break them down.
- The written word is powerful. Use the right methodology to get your message across, and make it appropriate for the person you wish to address.
- Adjust your voice and method of delivery on the telephone to create the right impression.
- You can build rapport with others by considering body language and the use of wording and the impact it can have on others.
- Above all, actively listen to gain a real understanding of what the other person wishes to communicate.

Chapter 13

Breaking down inter-team/departmental barriers

This chapter will explore the concept of the internal customer and will give you a seven-step model to aid you in clarifying their needs and requirements. To be an effective team leader you must be able to network, to influence colleagues at different levels within your organisation as well as customers outside. You will be shown how to create this 'web of influence' in order to gain resources, budget or simply the help you might need to finish a task.

13.1 Internal customers: who they are and their rights and responsibilities

The success of your team is dependent on each member being accountable. If you think of a football team, when an individual player is interviewed at the end of a game, he will always stress the importance of the contribution of others when he scored that wonder goal. It is no different in the business world; however, a team in the business context is not only reliant on the accountability of its own members, but also on how well it relates to other teams or departments. If you work in any part of a business and are constantly in disagreement with another department then it is going to be challenging to complete projects and assignments on time and within budget.

Unfortunately, the 'silo' mentality in organisations is prevalent. Many companies, especially American ones (for example, Marriott International and Mars), have taken a step towards overcoming the mental block that often precludes the perception of colleagues as 'customers', by describing staff as *associates*. This is hardly a revolutionary step (more evolutionary), but the word 'associate' has a connotation of a person being a part of a company (which 'employees' are) yet apart from it (which 'employees' are not), and this separateness lends a feeling that peers and colleagues should be accorded the same respect that customers (also separate from the company) are shown.

So, if you think of those you link with within your organisation as 'internal customers', how should you relate to them to get the cooperation that you need to be effective?

There needs to be a balancing act. Both parties have certain rights and responsibilities.

13.2   The rights and responsibilities of colleagues

13.2.1   Rights

Internal customers have a right to expect and receive:

- courtesy and respect
- information germane to their tasks
- resources necessary for the completion of their tasks
- freedom and opportunity to express views and opinions which contribute to decisions affecting their work
- understanding, if a request of them is unreasonable or prejudicial to their personal beliefs or needs
- support to fulfil corporate and/or departmental objectives
- honest, ethical, moral and legal conduct from colleagues and superiors

These rights derive not only from basic human concerns but also from an individual's role in the service chain.

13.2.2   Responsibilities

These are the other side of the same coin, and therefore include:

- being available, approachable and responsive
- listening fairly and uncritically to others' views and opinions
- sharing decision making when this affects others' work
- providing requested information and resources
- being aware of others' personal beliefs and needs and taking account of them when making requests
- honouring these responsibilities by being 'other-person centred'

13.2.3   Internal relationships

Respecting another person's rights and meeting one's responsibilities stem obviously from an attitude of mind. However hackneyed it sounds, it remains true that every service chain is wholly dependent on the interactions between internal people. Without their cooperation and collective dedication to service, the ultimate people, *external* customers, would be poorly served.

Case study: Adidas

The notion of sharing expertise, networking and the creation of internal champions of cultural change is critical within Adidas. The focus of collaboration is built into their competency framework, and mentoring and coaching are actively encouraged at all levels.

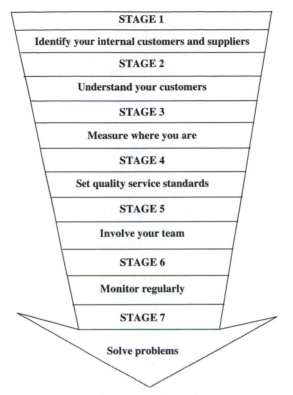

Figure 13.1   The seven-stage process to good internal customer care

Getting internal relationships right is obviously highly important. A seven-stage process to creating and maintaining the necessary quality of interrelationship will help engender the cooperation and collective dedication referred to above. Figure 13.1 shows in diagrammatic form a very practical approach to breaking down barriers and getting to know your internal customers.

Exercise: Seven stages to internal customer care

Stage	To be done
1 *Identify your internal customers and suppliers*	• Consider the work of your team or department. • Identify groups, departments and individuals who depend on your work. • Consider regular and irregular contacts.

2 *Understand your customers*	• Talk to them. • Find out what they want/need. • Identify problems caused by your work. • Consider the type of service you provide, i.e. 　– orientated to people or systems 　– high-tech or low-tech 　– level of complexity 　– how much contact is there with the customers?
3 *Measure where you are*	• Carry out an 'audit' or assessment of current provision. • Take feedback from customers. • Look for procedural bottlenecks and communication gaps. • Identify regular internal grumbles. • Examine complaints from external customers. • Identify routine failure to meet standards.
4 *Set quality service standards*	• Consider two dimensions of standard: 　– procedural 　– personal • Agree areas where standards can be set. Examples include for procedural standards: 　– time 　– workflow 　– flexibility 　– anticipation 　– communication 　– customer feedback 　– organisation 　– supervision • For personal standards: 　– appearance 　– non-verbal skills 　– attentiveness/interest 　– tact/patience 　– guidance/support 　– problem solving
5 *Involve your team*	• Accept you can't do it alone. • Clarify responsibilities for customer service. • Train staff if necessary. • Discuss/share new ideas and standards.

	• Involve everyone in setting new standards. • Ask for and use good ideas. • **Sell** the concept.
6 *Solve problems*	• Set up three simple feedback systems. • Service audit system (should focus on key indicators of quality service). • Internal customer feedback system. • Team feedback system. • Set up systems which concentrate on what's going well. • Involve your team in design of the systems.
7 *Solve problems*	• Accept that problems will occur. • Let team members know that mistakes are acceptable. • Avoid blaming. • Accept complaints and problems as ways of improving. • Make it easy for internal customers to complain. • Set up simple, understood procedures for dealing with problems and complaints. • Analyse problems to prevent recurrence. • If appropriate set up a project team to focus on and solve persistent problems. • Train team members to take internal complaints and problems seriously.

13.3   Influencing your internal customers

As you can see above and can experience in your everyday working life, there are many ways in which you need to influence others in your working environment. It could be your boss (during training programmes we usually find this is where most people want to focus!), your peers, people in other teams or your reports.

What do we mean by influence? One definition says it is the process by which you persuade others to follow your advice or suggestion. We like to define it as *the act or power of producing an effect without any apparent exertion of force or direct exercise of command.*

It is:

• getting what you need and/or want
• sustaining and/or enhancing the relationship

It is not:

- manipulation
- coercion
- command and control

There are formal processes by which you can have an impact on others. One way could be through the creation of a cross-functional team. This might be for developing a new product or the introduction of a new service you are offering to a client. It could be for an ongoing problem-solving process or quality improvement activities.

Equally well, you might need to influence a person on a one-to-one basis using an effective communication methodology or use interpersonal skills for mustering help from colleagues.

So let's look at each of these approaches.

13.3.1   The formal approach

There is a typical situation engineers have to face, that of sales staff in the organisation selling a new product or service to clients without bearing in mind the consequence that this will have on engineers and their workload. The engineers will not only have to maintain and support the current range of products/services but also be able to incorporate the additional work created by the new business.

How have 'best practice' organisations gone about getting over this problem?

Case study[1]

In the Spring of 2008, the Design Council in the UK undertook extensive research into eleven of the world's leading companies, including BT, Yahoo!, Starbucks, Lego, Sony, Virgin Airlines and Microsoft. The aim of this project was to offer information and ideas to other companies to help them strengthen their brands and gain competitive advantage in their own fields of business.

This research focused specifically on the design element, and innovation was one of the core brand values in all of the companies that made them 'stand out from the crowd'. Interestingly though, and relevant to the problem indicated above, the other factor that the research highlighted was the level of cooperation and interaction between designers and people working in other parts of the business, including engineering and manufacturing. At Lego (the world's sixth-largest toymaker), for example, the designers have to be able to speak fluently about the commercial implications of their design decisions. At Xerox designers are well versed in the analysis methods and processes used by their engineering colleagues. At Microsoft, at the development stage of new products or services, they put together a cross-functional team of user-researchers, engineers and product managers, as well as designers at the initial design stage.

It was a not just the fact that cross-functional teams were set up at an early stage that came through in the findings of the report, it was the fact that designers had shown a 'get up and go' attitude in seeking out ways of being involved in the wider business.

So, transferring this to your environment – are you being proactive in seeking out ways to be more involved at an early stage in the buying cycle, when sales staff are in initial discussions with the client about a new product or service? Do cross-functional teams get created at an early enough stage for you to have an impact? Tom Peters argues that 75 per cent of a middle managers' time should be spent on horizontal relationships to speed up relationships in the middle of the organisations.

Are you asking the right questions, flagging up your concerns about service levels? I'm not suggesting that you be a 'killjoy' to the sales staff, but make them aware of what is required when new business comes online. Sales staff are very conscious that if the after-sales service is not up to scratch, the client will bear this in mind during their next buying cycle. This could result in the client not even considering your organisation to pitch for new business.

Case study: Novellus[2]

Semiconductor equipment maker Novellus, based in San Jose in California, have a history of not only getting internal groups together at the design stage to form cross-functional teams, but also include their suppliers in these groups. They have been able to implement Lean production processes for new product releases instead of just existing ones, thanks to this involvement. This has led to sustainable competitive advantages, including improved credibility with customers due to shorter time to market, predictable completion dates, and successful launches.

13.3.2   Communication methodology

You are the vehicle for ideas from your team to reach senior management and in turn for strategic information to be passed on to your team. The effectiveness of your team is heavily dependent on your ability to influence your manager or superiors.

When you need to put a proposition together for approval – for example, the development of a new range of services – you must be rational in your explanation of ideas and present *benefits* to senior management. Your proposals or plans should be complete, not piecemeal, and if appropriate, use comparative or quantitative analysis to win their support. You might also use surveys, actual incidents or interviews with stakeholders.

If at first you don't succeed, don't give up! You need to be persistent. Ask for the support of other internal and external people who senior management will listen to. If you can get your hands on the information, show where the same idea has worked elsewhere and with what results. If you are successful in pushing through an idea or particularly challenging project you will gain respect from others. Success breeds success – and you and your team should get the better assignments in the future.

13.3.3   The interpersonal skills approach

Earlier in the book we looked at the key characteristics of a good team leader (see Chapter 3). From the spread of competencies that have been highlighted, the capacity to be a good listener and approachable to others are of particular importance in circumstances where the team leader needs to exercise influence. In *The Changing Face of Management*, Sinoo, Willenborg and Rozendaal[3] highlight the need for managers to demonstrate:

- strategic orientation and the ability to exercise strategic control
- a strong sense of practicality
- stimulation of action by creating constructive dissatisfaction
- strong intuition, sensitivity and willingness to experiment
- broad influence in the organisation and its surroundings
- willingness to listen, explain and offer assistance
- willingness to take risks, decisions and responsibility
- strong performance drive and exemplary behaviour that motivates and inspires employees
- guidance of employees rather than issuing orders
- authority through skill rather than hierarchical position
- use of events as a means of demonstrating and promoting policy

13.3.4   Getting the results you want

Given the above list it's clear that as team leader you have many tools which can be used to increase your influence and effectiveness. Some are not entirely under your control – such as word of mouth – though it is important to try and discover what others are saying about you. Seek feedback from others and use it to enhance your influence.

> 'To effectively communicate we must realise that we are all different in the way we perceive the world and use this understanding as a guide to our communication with others.'
>
> Anthony Robbins, American adviser to leaders[4]

Other tools are under your control. As a team leader, your role is no longer based on your knowledge and technical abilities alone. You need to build a 'web of influence' around you in order, for example, to be able to gain funds for projects/resources, or to ensure that the right people join your team. (You also need to have a 'sponsor' at a more senior level – we call this 'the shotgun rider' role – to support your activities).

So what do we mean by a 'web of influence'? People throughout your organisation and even colleagues in other companies can all make a contribution to your working activities, by sharing knowledge, informal advice, 'political' information about who are the real power brokers, etc. You need to invest time and effort in developing these networks. Obviously a mentor can help you in certain situations, but there are a whole range of areas when you need to call on others for help. This can be as simple as calling in a favour, e.g. asking a colleague from another team help you put together the information packs for a conference, through to chatting about a situation that has occurred where you believe undue duress has been put on you, and which in your eyes has moved into bullying or harassment. What should you do? How should you go about handling this?

If you are calling on favours from people there needs to be a degree of give and take in the situation. You must support them in informal ways as well; it can't be a 'one-way passage'.

How do you build a network of contacts?

- Volunteer to be part of a committee.
- Secondment.
- Ask to visit another department or team for the day to see how they operate.
- Go out with a member of the sales team for a client visit, and gain a better understanding of what they have to face.
- Invite people from other teams to come into your team for the day to see the issues your group have to face.
- Develop relationships with decision makers in other industry sectors, either in a comparable role to yours, or at a more senior level.

The 'ambassador' role of getting both yourself and your team known throughout the organisation is vital.

Influence = better results

On the whole, all of us want to be successful. This is in part the simple, natural desire to do well and earn the esteem of our colleagues and superiors, to achieve targets and the bonus cheques that go with them. In part it is a feature of the special psychological characteristics which makes good team leaders, 'people' people. Fundamentally you need to be interested in people and what motivates them. If you are not, why should they be interested in you? Good influencing is all about 'inspiring people to act according to their own motivations'. A successful, influential team leader is one who inspires a lot of people.

It follows that to obtain better results, the team leader has to have a way of being in touch with a broad cross-section of colleagues, doing whatever it takes to make them more likely to say 'yes' to your ideas, requests or suggestions.

So how can you do this?

The message that precedes you

When you walk into a supermarket you are faced with thousands of different kinds of foodstuff, along with a selection of personal or household items. You are faced with two types of choice:

- Which of maybe up to a dozen brands of some staple item should you choose?
- Should you buy something new or not?

There is no-one to guide you. All you have is the label. So how do you choose? For many items you choose the same label you bought last time you were there, but there is a first time for all of us, and our first choice is made either at random, on price or because of a message you received before you went shopping. For foodstuffs bought in supermarkets the primary 'influencer' is advertising.

Advertisements are often backed up by personal recommendations or warnings, (for example, Britain's consumption of cranberries a few years ago broke all records as a result of a series of recipes by a well-known TV cook), and reviews in newspapers, magazines or consumer-oriented TV programmes. Advertisements, of course, can be undermined by other information, such as:

- Press comment relating to prosecutions for poor hygiene or reports of sabotage (contaminants in Brand X products).
- Press comment on poor personnel practices or falling market share (we like to 'back winners').

All of these influences work in exactly the same way in the business world. If your organisation's marketing and PR departments do a good job, people who buy your organisation's products and services will be half-convinced before they buy. If they do a poor job, they make it harder.

So how does this relate to you as a team leader and your capacity to influence others?

Enlarging your 'web of influence'

There is an invisible message that precedes you when you deal with colleagues. Do they find you good to work with, or a challenge? So, looking at enlarging your 'web of influence', the aim is to increase the number of people who 'promote' or 'sell' you to others on your behalf. They are active referees who speak in a positive way about you. How many people do you work with who act as a reference for you? If you don't know, change this immediately by asking your internal and external customers the following four questions:

1. What do you like about working with me/my team's work?
2. Where do you see room for improvement ?
3. What do you suggest we do about this?
4. Agree an action plan.

Personal recommendations from satisfied managers, colleagues and clients are most effective since they involve your personal reputation as a highly competent team leader.

Organisational decisions are influenced by commentary from others. This may be unprompted, as when a manager happily talks about you or your team, or prompted when others are asked questions about you. 'Nothing gets done till Peter S gets involved. Nothing is too much trouble for that guy.'

Of course, unhappy referees will be only too pleased to ensure that everyone they meet knows of both their misery and who caused it: 'one satisfied internal customer brings in four others; one who is dissatisfied chases ten away'.

A group of happy references prepared to recommend you and your team acts as a band of 'assistants' and clearly extends your circle of effectiveness. As your band of 'assistants' grows, you should observe a 'ripple effect' as your circle of effectiveness widens.

Pat saw this in action when she had a conversation with a management development consultant colleague of hers, Mike Kearsley, who worked for several years in a leading accountancy firm in London. He described how one of the partners underestimated the importance of clarifying and promoting the services he offered to internal colleagues.

Case study: Influence in a leading accountancy firm in London

The VAT partner gained all his work through the other partners. This work was thin and sporadic and his future was uncertain. I asked the others why they didn't involve this expert and it was because they felt he highlighted their mistakes, was an extra fee and clients had to pay more VAT. He was asked to think of times he had saved clients money and wrote out fourteen examples. The first was where a client owed £4 million, was being taken to court by Customs and Excise, and faced a jail sentence. The partner worked on his case for a few days, got him off the court case and charges and also got him off all of the £4 million for a very nominal fee. When this story and some of the others were sent to the partners – one each week – work began to flow in. Why had he not told the others about this? He thought it was being too pushy and like a salesman!'

Help others

Active references are most likely to agree to take on 'assistant' status if you help them to achieve their own objectives. In summary, you help them to satisfy five desires:

- to be secure
- to be noticed (in a favourable light) by colleagues and friends
- to have a sensation of wellbeing

- to possess something new (being up to date or a little ahead of the pack)
- to be approachable and helpful to their colleagues

You do this by keeping your conversations centred on the other person, since they are usually pre-occupied with their own problems, motivations and desires. When you meet your internal customer colleagues ask open-ended questions, listen carefully, explore negatives about yourself or your team with a view to either neutralising them or turning them to your or their advantage. Stay positive in attitude and show them esteem.

You can also increase your influence by taking every opportunity to be helpful. Your primary 'tool' in this area is very simple: maintain the relationship. We have probably all experienced disappointment when an apparently close friend moves out of our immediate circle and then fails to keep in touch. That single 'missing' greetings card on our birthday or at Christmas causes more pain than the pleasure we get from a fistful that arrive from people we see regularly. The lesson is clear, relationships, like cars, need regular servicing!

For internal customers this means dropping them a friendly email occasionally, being prepared to support them if they need help in some way, and letting them know of any information you might come across that could be pertinent to their team or department. If your colleague is in the design department, this could be flagging information about a competitor bringing out a new product. If they are in research, new funding schemes that you might hear about, etc.

If you work as a sales engineer, this means that you don't simply drop out of your client's life as soon as he or she signs the contract. It may well be that other people take over day-to-day responsibility for delivery, installation and maintenance of your product, but you should retain responsibility for maintenance of your relationship with your client. This can be done by a judicious mixture of formal and informal contacts and meetings.

Formal contacts should be arranged on a regular basis at a frequency dictated by the circumstances. The agenda for these meetings might include:

- Your expressions of thanks for any recommendations they may have passed on, and the beneficial impact of their contribution, if any, then:
- Is the business case for your product being satisfied?
- Speaking now from experience, how does your or your team's solution to their business problem measure up?
- Is there anything you can do or arrange to improve the client's experience of your product or service?
- Are there any related business problems which could solved by extending the existing solution (you are, after all, a sales engineer, always on the look-out for new business) and new active references?
- Are there any other issues that have arisen since you last spoke, where you might be able to help?

The primary talking point, as always client centred, is their experience of you and your team and how that experience can be enhanced. The secondary points are probes to discover new ways in which you can improve their performance and reinforce the warm feelings that their initial purchase gave them.

Informal contacts are not necessarily face to face. You might telephone with a bit of urgent news or information, or write a pleasant note to accompany written material. You will ensure that your client is always invited to appropriate company-sponsored social occasions: the Christmas drinks party, the golf day, as well as your pavilion at trade fairs or at sporting or cultural events. You will ensure that he or she is on the mailing list for company client magazines, Christmas cards and other team update mail-shots.

Once you have secured a contract, you should know quite a lot about your client and his or her particular interests, both in and out of the workplace. Send them clippings from magazines and newspapers and any other oddments which come your way and which might interest them. You may have ideas about how they can extend their areas of influence. Perhaps on a trip abroad you spot a market or some other opportunity they could exploit; you might come across information about their competitors which could be of interest to them. Each sending is a reminder that you care and that they are special to you. This means there is more of a chance that your client will respond in kind, and automatically continue to do business with you.

Here is another case example from our colleague Mike Kearsley.

Case study: Working with professional firms throughout the UK in a short-term development programme

Each group consisted of some eight or ten tax managers working with a partner. They agreed to meet for breakfast every two weeks for three months and then each have a short personal chat with me as the facilitator throughout the same day. They set small targets for a single activity during each day – perhaps one call they would not have made or an extra question during a meeting. At the core were the questions, 'How is business generally?', 'What is your major problem at the moment?', etc. They were to spend no more than ten minutes with this each day, keep a record and report what had happened each fortnight. As a result it was not uncommon for teams to uncover as much as £100,000 of new business which they would previously have ignored. All this through talking to their clients a little more, talking to their colleagues a little more and having a short plan.

Attitude

Of course, the one thing you do not want to do is come across as insincere. This comes down to a mindset. Put yourself in your internal customer's shoes and think 'what can I do to help them achieve their objectives?' By aiming to create a professional friendship and a potential referral first, and influence your colleagues second, then you do not come as an aggressor, but as a potential ally.

Much the same process can be done with the 'personal chemistry' between you and others. When you first meet people, you gain a first impression which enables you to slot them into a scale, running from 'love at first sight' to 'mad, bad and dangerous to know'. You also know that the extremes are artificial: there are people you detest who manage to have warm relationships with others; they cannot be all bad. Similarly, many of your best friends may be openly detested by others of your acquaintance, so they are perhaps not as perfect as you see them.

The truth is that we all have strengths and weaknesses, and we look at other people through the distorting lens of our own prejudices. For example, a person trained from childhood always to have gleaming shoes 'converts' the first impression of someone with scruffy shoes into a negative, then magnifies the distortion; he or she is unkempt and therefore lacks self-discipline, and is probably a scoundrel and best avoided. In fact, of course, you are not a scoundrel; you simply had to walk across a muddy car park and forgot to brush the mud of your shoes before you came in to meet your colleague! Since in business meetings you act to some degree behind an 'official' veneer, you have an opportunity to overcome your colleague's negative first impression. You must engage your self-confidence and positive attitude, show your good points and demonstrate that they outweigh the initial negative.

We suggested earlier that your first priority should be to develop professional friendships and try to gain an 'assistant'. If you keep that in mind you will be able to neutralise most 'bad vibes', and at least gain respect for your helpfulness and professionalism, even if the relationship never has much personal warmth. If the other person is prepared to pass on an acknowledgement of that respect, then he or she will be an (albeit unenthusiastic) 'assistant'.

The opposite side to this argument also applies: even if you gain an unfavourable first impression of others, they must have some good points. You must always seek to see them in a positive light, so that you genuinely want to help them achieve their goals. And with this positive attitude you should succeed.

You reap what you sow

Influencers are special people. They have a particular interest in and aptitude for seeing the best in people. Liking people is as much part of their stock-in-trade as their technical skills. This gives them their optimism and positive attitude, which makes it relatively easy to develop a friendly relationship with others and turn them into willing 'assistants' or active references.

Above all else seek out like-minded people in your organisation, so that you can champion and aid each other. It can be pretty isolating trying to do everything on your own!

13.4 Chapter summary

- Identify your 'internal customers', understand their needs and requirements, put a measuring methodology in place, set quality service standards and create a system of feedback to ensure continuity and development of the relationships.
- Remember that the key to being an effective influencer is to be genuinely interested in others; put yourself in their shoes and help them achieve their objectives.
- You and your team need help from others, from within and outside your organisation. Develop a network of like-minded colleagues to be mutually supportive of each other.

12.4 Chapter summary

Chapter 14
Handling conflict

As we saw in Chapter 1, Tuckman talks about the 'process of enjoining' and describes four stages of team development.[1] Groups and teams must experience various developmental stages before they become fully productive and function as a group or team entity. Chapter 1 stated that for optimum group performance all four stages – forming, storming, norming and performing – need to be negotiated. The group must not get stuck in one phase. And this is where conflict comes in! The goal of conflict is getting to **performing** and **growing** the team. Conflict is normal. The team leader must create an environment where everyone can express themselves honestly. This chapter will help you not to get stuck in a conflict or 'storming' situation and be unable to get out of it.

Handling conflict is really straightforward. Believe us! It is an opportunity for you to resolve a situation or conflict once and for all. Most team leaders (and others) do not relish handling conflict situations. What happens? The situation only gets worse! The team leader who wants to do his or her job properly, who wants to control change, must relish – *yes, relish* – handling conflict situations. The point of this chapter is to explain to you how this can be realised.

14.1 'Yes' to the person and 'no' to the conflict

How do you say '**yes**' to the person and '**no**' to the fact, the problem or the conflict? Another way of putting this is how can you bring yourself from the NO situation to the YES situation? This is done by finding a common viewpoint! Team leaders who are a positive example in controlling change handle a conflict somewhere between rules and arbitration. This means that handling a conflict means handling power and knowing how much power you possess in any situation.

The questions to resolve conflict focus on:

- defining causes
- understanding points of view

- clarifying preferences
- examining alternatives
- gaining commitments from those involved that they are prepared to work towards finding solutions

The fundamental ways to find joint solutions to conflict include the following:

1. Acknowledgement.
    Listening is the first sign of respect for the other person.
2. **Yes** to the principle.
    Get a 'yes' to a known rule, policy or principle. Never be personal!
    - 'Do you agree that our working hours are from 08.00 to 17.00?' (known rule or policy)
    - 'Do you a understand that all documents market – 'private' should never leave the company premises?' (known rule or policy)
    - 'Do you accept that our organisation's success comes from giving customers, if not more, at least what they expect they are buying?' (Understood principle)
3. **Yes** to the situation.
    The first yes after a number of **no's**!
    - 'Do you agree that we are in conflict?'
    - 'Do you agree that we have a problem?'
4. **Yes** to commitment. Looking into the future!
    - 'Are you prepared to do everything possible to solve the problem?'
5. **Get straight** to the point by recognising and estimating the forces in play.
    - 'This is your **power or authority. You can do this or that'**
    - 'This is my **power or authority. I, on my part, could do this or that'**
6. 'What do you propose we do to resolve this situation?'
    - 'Where is our halfway point?'
7. To jointly set concrete goals.

Within all organisations, conflict can be resolved at three different levels:

1. Mutual recognition of a known rule or policy or principle and agreement on its application. Of course, the rule or policy must exist and be familiar to everyone. Refer to your organisation's HR or quality handbook or company rules.
2. In the case of the principle you must ask a closed question to get a 'yes' response.
    - 'Do you agree that our success comes from putting our best personnel in positions where they'll have most impact?'
    - 'Do you accept that we should treat others the way we ourselves would like to be treated?'
    - 'Do you agree that when we make a promise we should do everything to deliver it?'

Never be personal. For example, never say 'Do you agree that you were late for work?' Say, instead, 'Do you agree that our working hours are from X to X?'

3. Mutual agreement before things get out of control. For this to happen requires a dialogue between 'adults' who are both willing to find a solution.

If none of these options are possible there is only one other logical option: arbitration or some higher authority.

Case study

You are a sales engineer:

Yesterday, your management decided to launch a new product. You see this decision as a personal victory and a new opportunity for your department. Thanks to your encouragement, all your team members are enthusiastic and impatient to start marketing this product.

You have committed yourself to supplying the product for them to sell in a month's time.

You know that the technical manager is still not happy about the quality of this product, but s/he will just have to 'get their act together'.

You are the technical manager:

Yesterday, management decided to launch a new product. You feel that this is a mistake as any product launch is way too premature. Your production department is already overloaded and more extra work could have adverse consequences for the quality of your products. Your team share your concern and you have promised to ensure that nothing new will be started in the next three or four months.

You are familiar with the sales engineer involved and his aggressive attitude. You have decided that in this case, product quality cannot be jeopardised by overhasty action. The company's and your department's reputation is at stake.

Case solution

Technical manager to sales engineer:

1. 'Hello Paul, I've asked for this meeting as there's something very important I want to discuss with you. I believe it's of real importance to the wellbeing of the company, our customers and my department.'

Sales engineer to technical manager:

2. 'What do you have in mind?'

Technical manager to sales engineer:

3. 'Do you agree that we must do everything possible to preserve our product quality and deliver to customers products that actually do what we promise they'll do?'

→ Get **Yes** from the sales engineer.

Technical manager to sales engineer:

4. The first yes after a number of **no's**!
   - 'Do you agree that we have a problem about the launch date of the new product?'

Technical manager to sales engineer:

5. **Yes** to commitment. Looking into the future!
   - 'Are you prepared to do everything possible to solve our disagreement are reach a win−win solution?'

→ Get **Yes** from the sales engineer.

Technical manager to sales engineer:

6. **Get straight** to the point by recognising and estimating the forces in play.
   - 'This is your **power or authority. You can do this or that**.'
   - 'This is my **power or authority. I, on my part, could do this or that**.'

Technical manager to sales engineer:

7. 'What do you propose we do to resolve this situation (and reach a win−win)?'
   - 'Where is our halfway point?'
8. To jointly set concrete goals.

14.2   Other examples of conflict

Every discipline of football may be viewed as a form of ritual combat, with opposing teams using skills and guile to score points. For the most part the violence and aggression are under control, but occasionally players allow competitiveness to degenerate into conflict, and succumb to the temptation to perpetrate fouls or throw punches.

Workplace conflicts are less exciting to watch though they can, in their own way, have the potential to be equally damaging. For example, see the case below when a sales manager approaches a stores manager demanding immediate delivery of a vital order, perhaps to a new customer who could become an important long-term source of business.

'Everyone knows that our stocks of that product are fully committed − if this order were so important, you should have gone through the proper procedures so we had time to organise a new supply.'
'We telephoned you as soon as we signed the deal. You must be able to do something and, anyway, customers are more important than your paperwork.'

'I'm sorry. We don't have the stock so there's nothing we can do. You'll have to tell your customer to wait. This isn't the first time your gang have ignored the procedures and expected me to work miracles. What is it about you sales people?'

And off they go, like kids in the playground.

Our newspapers are full of similar situations in other circumstances: acrimonious divorces which end up in the courts or a prison cell, friendly neighbours who fall out over untrimmed hedges, un-trained dogs, rights of way, noisy teenagers, self-help groups who descend into violent vigilantism, and so on up to civil wars and global conflict.

We will now look at some of the causes of conflict and ways to avoid it.

14.3   The mechanics of conflict

Put simply, we get a loud 'bang' whenever an unstoppable force meets an immovable object. However, the people in the above scenarios used to be, and probably still are, perfectly reasonable, normal human beings: so what caused them to 'flip'? The clues are in the workplace story immediately above:

- **Thoughtlessness**
  Newton's Law that every action has an equal and opposite reaction transfers quite easily into human affairs, and here the sales manager had become so excited about the new order that he or she had forgotten to think about the consequences. It wouldn't have been difficult to have called the stores department to say, 'I think we have a good chance of securing this order; when can we deliver?' before going on the sales call. However, whether or not the intention was there, the call didn't get made. Nor did there appear to be any serious attempt to ensure that success was communicated properly at the earliest possible moment. Thoughtlessness is the opposite of planning and preparation, and 'no surprises' is a sensible motto to carry through our business lives.
- **Frustration**
  The stores manager is perfectly aware that customers must be served, but has other priorities including minimising stock balances. Simple procedures have therefore been designed to minimise stock while maximising customer service, but the balance is fine, so that large, unexpected demands cannot be met. Everyone knows this, yet (some) sales people persist in ignoring the realities of running an inventory.

Maybe we should restate the first sentence in this section: Conflict occurs when thoughtlessness meets frustration.

- Other factors may of course be present. One or both of our protagonists might have a bad tooth, or air-conditioning stuck on 'Min' or 'Max', or some personal problem. Maybe the two of them simply don't get on. Ordinarily, petty irritations like these wouldn't much matter. They would be brushed aside with a joke, a state-ment of sympathy or hidden behind a veneer of professionalism: but they do shorten fuses, making conflict more likely.

So, if thoughtlessness and frustration provide the tinder for a full-scale row, what provides the match and fuels the conflagration? Read the exchange again, looking for little clues.

- Did you see that both parties were in a way actually seeking a fight? Did you notice the 'language' placed there to be read: *'The sales manager demanded immediate action'*?
- The stores manager says *'Everybody knows'* and *'What is it about you sales people?'*, implying that the sales team is staffed by thoughtless, pushy individuals with the planning horizons of fruit-flies.
- The sales manager denigrates the procedures, suggesting that the stores people are rule-bound simpletons without the imagination to rise to the challenge. The language used by both parties shows that they were egging one another on, rather than looking for ways to calming things down and finding a solution to a problem.

You know this from real-life arguments. You know when someone comes up to you whether they are likely to be confrontational or seeking consensus, and you have a choice. You can seek ways to deflect or smother their anger, or pitch in and encourage it. We sometimes actively contribute to starting a row to relieve internal anger and generally 'let off steam'. But we also know that having a good shout can move on to more serious things, and usually have the good sense to back down before perma-nent damage is done. Controlled conflict can, indeed, be helpful, as in our football example where it is the essence of the game; the boosted hormone levels enable players to perform miracles they could never achieve 'cold'.

14.4   Stopping it from starting

There are a number of proverbs such as 'Prevention is better than cure' and 'A stitch in time saves nine' which explicitly recognise that it is better not to get into a mess than to have to find a way out later.

In our present context, one party to a potential confrontation is often somewhat in the wrong and is in reality looking for help rather than a fight. In the world of business, we expect or desire that the professional relationship will continue, so it is wise to recognise fault or error for three reasons:

- It gives the other party the illusion of occupying the moral 'high ground', from which to bestow favours.
- It takes the wind out of the other party's sails, defusing the impulse to attack.
- By and large people like to be helpful, if only to clock up favours to be repaid later.

In our example the sales team were in error in not forewarning stores of a potential large and urgent order. So a hang-dog expression and an apology would have set an entirely different tone to the meeting, leading to a win–win outcome whereby, for example, the two teams might work together to arrange partial shipments to several customers including the new one, with the sales team picking up the bill for the secondary shipments. This could have far wider benefits, since several customers would be brought into the friendship act, helping to cement all of the trading relationships into a network of helpful, trusting partners.

In this case there might also have been a 'system' failure in that the ordering paperwork didn't make provision for forecasts of likely demand – something that might have been recognised towards the end of a friendly encounter, and rectified to prevent recurrence of the problem. This array of internal and external benefits is in stark contrast to our original scenario which ended with nothing resolved, two people firmly dug into entrenched positions and a pair of tarnished reputations.

In a wider context, conflict can often be avoided by keeping a sense of proportion. We all know that little things can itch and niggle and grow into major problems unless they are identified and dealt with early on. Gardeners will recognise that pernicious weeds such as ground elder or Japanese knotweed have to be destroyed immediately they are detected; otherwise they can take over the entire garden and you end up having to kill valued and favourite plants in order to eliminate them. Our reactions to mildly irritating personal habits in colleagues or partners can balloon out of control, completely over-shadowing the reasons why they were welcomed into our circle in the first place. This can leading to irreversible relationship breakdowns unless they are discussed and brought into perspective at an early stage.

So too in business. The team leader's primary task is to help ensure the long-term survival of the enterprise; this usually involves building long-term relationships with trading partners within a context of good service, good products and good reputation. So anything which undermines this context is bad and must be avoided.

We all make mistakes; we all do rather silly things from time to time. Sometimes we forget to do things and we may fully deserve the occasional reprimand. However, when a problem arises the first thing to keep in mind is the health of the enterprise and how the problem can be solved or its consequences minimised or deflected. Blame and retribution (or, better, calm analysis and avoidance measures to prevent future repetitions) must be kept for later. So impress upon yourself and those around you that when a problem arises:

- Analyse the problem not the people (it may not be the first suspect's fault).
- Protect the trading partners (if only to give them the earliest possible warning of non-performance so they can take avoiding action).
- Learn from the experience (get something positive out of potentially negative situations).

One specific benefit of this approach is that the 'post mortem' is taken away from the immediate proximity (in time and, probably, place) of the source of potential conflict. All parties will have time to calm down and they will have to cooperate to mend the damage. This automatically restricts the opportunity for conflict rather than consensual resolution of differences.

14.5   Diverting the avalanche

No matter how hard we try to apply the suggestions of the previous section and avoid conflict altogether, sometimes the little reasons, the toothache, the humidity, and now some thoughtless foolishness bring us to the boil. Wittingly or otherwise, we get into conflict. What do we do now?

The first thing to recognise is that there are several kinds of conflict. The least serious might be called 'pseudo-conflict' – the two parties have no serious problems and are really just letting off heads of steam built up over a period as a result of myriad minor frustrations, and some trivial difference or interpretation of body language produces the explosion. The first one of you to recognise this can allow the other to run out of steam and then work to defuse the situation with a joke or change of subject, and you can resume a friendly relationship again.

Proper conflicts arise over serious differences, deeply held beliefs and territory (real or imagined). Our news media are full of real-world conflicts of this type, and it is clear that in the more extreme cases the best efforts of top arbitrators are unable to bring resolution. Sometimes the 'boss' (in the form of the UN) steps in to impose a solution, but even this works only when the protagonists truly want peace and reconciliation.

Workplace conflicts, of course, rarely reach fisticuffs, though the same three potential solutions remain available:

- Sort it out between yourselves.
- Call in an arbitrator.
- Appeal to a higher authority.

Let us look at these in turn, bearing in mind that the descent from different viewpoints through disagreement to an argument is a bit like an avalanche – far easier to stop or divert in its earliest stages. As it gains momentum the opportunity and likelihood of stopping the escalation of conflict without outside intervention reduces rapidly, eventually reaching a critical point where the only option is to allow it to run its course and then pick up the pieces later, when calm has been restored.

14.5.1   Sorting it out together

The first step in diverting the avalanche is for one participant simply to stop making things worse. 'Count up to ten' before making your next contribution, using the time to calm yourself a bit and think about what you are about to say. Remember that someone in full flood isn't necessarily going to be able to pull up immediately, so the initial procedure is to stop the escalation and stabilise the situation:

- DON'T rise to the bait – just agree with the other person.
- DON'T raise your voice – the other party has to 'change gear' and listen more carefully if you speak quietly.

- DON'T provide ammunition – copy politicians and speak without saying anything.

Eventually the other person will realise that he or she is pushing against air, and the conflict will peter out since there is no longer any resistance to fight against.

At this point you can start working towards a resolution of the problem (even if in the end you have to agree to disagree about the solution). The principle is quite simple: you must acknowledge that there is a problem, define the problem in objective terms (analyse the problem, not the people), and then try to figure out how to solve it.

'I think we have a problem' encapsulates this. It doesn't pin blame on the other party, but acknowledges that something is wrong and implicitly accepts at least equal responsibility. Note that the answer, 'You have a problem; everything is OK from my end' is not a good idea! Both parties have to agree that there is an issue to be resolved. You can then turn from fighting over it and 'hold hands' while cooperating in the search for a solution. Note that this must be a cooperative enterprise. If one party accepts responsibility for solving the problem while the other walks away then it is most likely that the conflict has only been delayed, while what you need is to remove the causes permanently.

It is essential to avoid returning to a conflict situation during this second stage, and a good way to do this is to ensure that every question is centred on 'What?', 'When?' and 'How much?', and avoids 'Why?' and, in particular, 'Who?'. Once the problem has been defined and agreed, some solutions are often obvious; you can work together to choose the 'best' according to whatever criteria seem relevant. This should if possible involve both parties, so as to reinforce their present and future cooperation as well as to ensure that both gain a full understanding of the effort involved, and become less thoughtless in future. This is an adaptation of the process of involving young delinquents in making good the results of their crime, with a view to having them understand the full consequences of their actions.

We are now back in a familiar planning and decision-making situation, where the normal rules apply: agree what has to be done, by whom and by when, concrete steps which can be monitored and controlled.

Close analysis often shows that there are actually two or more problems, and the cause of conflict is the fact that they have become entangled, rather than the problems themselves. Thus in our sales versus stores example, satisfying a customer order is one problem and fixing the system to prevent its re-occurrence is another. Even if the paperwork had been done correctly there might still have been a low stock problem, perhaps to be solved the same way (ringing around other customers to find some who will accept part-shipments), but the excuse for the conflict would have been greatly reduced.

14.5.2   Bringing in an umpire

The previous scenario described situations where the solution was within the authority and competence of the protagonists, and where they were able to divert the conflict before any significant harm had been done. This is sometimes not possible, for all manner of reasons including availability of resources, lack of competence or extent of authority. Appeal must then be made to a higher authority who can either impose a solution – really the last resort – or act as (or appoint) an arbitrator. For large-scale serious conflicts professional arbitrators are available, such as the Advisory, Conciliation and Arbitration Service (ACAS), best known for their work in mediating difficult relationships between managers and workforces, or judicial arbitrators operating as an alternative to full-scale civil court proceedings, particularly in the insurance and commercial fields.

While these high-powered umpires are outside the remit of this book, it is instructive to consider the advantages of arbitration over court action, since they apply at all levels:

- Arbitration is quick. The parties can get on with it without delay, the process is often quite short and the time and location of the meeting(s) can be fixed by mutual agreement.
- Arbitration is friendly. Unlike the confrontational atmosphere of a court, in arbitration people of goodwill are seeking agreement.
- Arbitration is private. The protagonists and the umpire hear all the arguments, but the potential for disclosure of sensitive or embarrassing information is minimal compared with a 'public' court hearing.
- Arbitration is final. Parties go to arbitration in the full knowledge and acceptance of the fact that, even if they don't like the outcome, there is no scope for appeal.

In choosing an arbitrator or umpire the primary qualification is that he or she is completely impartial and trusted by the disputing parties, as well as senior enough and of sufficient authority to command respect and be immune from browbeating. Depending on the circumstances you might choose a team leader/manager from another part of the organisation, or perhaps a non-executive director or a partner from the organisation's audit company or legal advisers. It may help if the arbitrator has prior knowledge of the subject to be discussed, but this is not essential since he or she can always call in 'expert witnesses' or ask for background documentation. Note in this context that the process of arbitration is not the imposition of a solution. It is the development of an agreement between the disputing parties.

In practice most of us act as arbitrators. Parents arbitrate disputes between their children; you arbitrate disputes amongst your friends and colleagues, and from time to time if a conflict is irresolvable you may need to take on an arbitration role. How is it done?

The purpose of the exercise is to reach an agreement, so the arbitrator's primary tool is listening and questioning to understand all the facts of the case. By active listening the arbitrator demonstrates respect for both parties and their viewpoints and increases his or her impartial credibility. As ever, analyse problems not people: 'What?', 'When?', 'How much?', rather than 'Who?', and continue questioning and analysing until each party agrees your statement of their case. Sometimes this may be done when everyone is present; on other occasions it may be more convenient to call in the parties concerned one at a time.

At this point the facts have been established. Now try to represent the problem as a sort of target. The outer ring is where both parties are in complete agreement, surrounding areas of minor disagreement but small importance, and the hard kernel of dissent in the centre. First work on the outer ring, getting both parties to say that they agree, and congratulating them on this albeit modest progress. The moment they both say 'Yes', they are beginning to cooperate to solve the central problem, which has at the same time been made a little smaller and more clearly defined. This is an iterative process, like peeling an onion, where each layer is smaller though perhaps more difficult to remove. Keep asking questions with a view to agreeing that this layer is either relatively unimportant as an issue, or perhaps a real issue but not of immediate relevance (perhaps it can be dealt with at another time), and pausing to offer congratulations and a summary of progress each time another step has been achieved.

As you penetrate towards the core issues it may be worth reminding everyone that the objective is to solve a business problem in line with corporate goals, not to score points off one another. A problem is solved, but no-one 'wins' or 'loses' (however they may choose to present the outcome when they return to their teams or departments).

The arbitrator should reinforce this by continually asking the parties to suggest solutions, beginning at the earliest possible stage, so that they become used to working together to solve small differences and establish a pattern which will later be used to address bigger issues. Ideally all of the initiatives should come from the people concerned in the dispute, though when they are stuck the arbitrator may be able to make a suggestion which breaks a threatened deadlock.

Reaching the core

Sometimes you will find that there is nothing there! The original 'problem' may have been a tangle of small, unrelated issues which together caused conflict, but the patient process of teasing out and discussing them item by item finally revealed the original issue to have been an illusion. Often there is indeed a real core problem, but by the time the surrounding 'noise' has been stripped away, it may be revealed to be a small thing, easily resolved now that goodwill and a hunger for agreement have been established. Or there may be a substantive issue, in which case, once both parties have made their suggestions, the arbitrator can take one of three paths:

- Find in favour of one side or the other.
- 'Split the difference.'
- Suggest a completely different approach involving a third party.

Now remember that at the outset the parties agreed to arbitration, and that part of the nature of arbitration is that there is normally no appeals procedure. So the final step is for the parties to agree on the outcome and to commit to one another and to the umpire to make it work. In simple cases they can shake hands and get back to work. In complex or sensitive cases it may be sensible to extend the arbitration to include working out the steps making up the solution, with concrete objectives and timescales and perhaps some review meetings where the arbitrator and protagonists come together to examine progress and ensure that the agreed solution is workable and doesn't have any unforeseen and undesirable side-effects.

The primary characteristic of arbitration is that, under the supervision of an independent umpire, the two parties cooperate to define, clarify and resolve the cause of the initial conflict. It is a mediated form of what we referred to as 'holding hands' in the previous section, centred on healing differences and establishing or re-establishing a friendly, cooperative working relationship for the future as well as solving the central business problem.

14.5.3   Calling in the boss

The solution of last resort is to call on higher authority. Our opening example of growing conflict between a pair of managers would probably have resulted in one or both appealing to more senior figures to impose a decision. This approach is fraught with difficulty:

* The authority figure is probably biased either personally in favour of one or other of the parties or by his or her business objectives (and was called upon for that very reason).
* There is a clear 'winner', and hence also a 'loser' who will in future be less cooperative towards the 'winner', so that the enterprise as a whole suffers in the short and medium term.
* It is too easy for busy executives to make quick but wrong decisions based on inadequate, incomplete or inaccurate information.
* There is a small, but real, danger that the outcome will cause the same issues to be escalated to a higher level in the organisation, rather than being solved.

A more senior figure will probably recognise these dangers, and will find time to arbitrate a solution, rather than simply listening to the arguments and imposing one. They could call in an independent arbitrator and in effect 'de-escalate' the problem back to the individuals concerned. Optionally, they could do both by making a short-term decision to get both parties to realise that actions are being taken to resolve the issues between them, while setting up the framework for a properly considered long-term solution.

14.6   Chapter summary

* The potential for conflict exists in every aspect of our lives.
* Prime causes are thoughtlessness (taking action without thinking about the consequences) and frustration where other people abuse our goodwill.
* 'Prevention is better than cure', so honesty in admitting fault or error is always best.

- Avoid conflict by keeping a sense of proportion: identify and deal with those little irritants before they grow into major problems.
- In conflict situations you must stop making things worse: acknowledge that there is a problem and decide what you will do to resolve it.
- You cannot resolve conflict on your own. If the other party or parties are unwilling to work towards a resolution, conflict will continue. Arbitration might be the only solution.
- In the above case consider arbitration.

Part 5

Career management

Chapter 15

Planning for the future

So your team is created and up and running. The structure, processes and team roles and responsibilities are understood, and the team has developed to the point where you can take a more 'hands off' approach. How should your own performance develop, now more of your time has been released from day-to-day operational activities, and what career progression might you be looking to explore?

15.1   How should you be spending your time?

Your focus should move from a tactical to a more strategic role. For example, you could capture and shape trends in your organisation's marketplace or you could display breadth of view across your entire organisation, external market and global perspective, taking a more medium- to long-term visionary view.

You should also be exploring ways in which you can enhance your own performance, your standing within your organisation and in the marketplace, and possible career development opportunities.

15.2   Improving your own performance

We were recently at a meeting in the City with Fraser Murray, Managing Director of Rock the Boat Consulting. He mentioned a new survey of the most successful organisational managers, one where they listed the top three things that helped them in their careers. These were:

1.   being good in your job or completing your assigned work successfully
2.   being a good networker
3.   having a mentor

Since this book has already looked in detail at point 1 of this list, let's move on to points 2 and 3.

15.3   Being a good networker

In Chapter 13, 'Breaking down inter-team/interdepartmental barriers', we have explored in detail the variety of approaches you can adopt to influence others and network with colleagues both within and outside your organisation.

Another way of creating a network is to join or establish an action learning set. Action learning is an educational process whereby those who participate study their own actions and what they have learnt in the process, in order to enhance their future performance. These learning sets can be created by people within your organisation who volunteer to join the group, or by participants from other organisations. In effect it is about understanding yourself, self-development and group-based learning.

Professor Reginald Revans, the originator of action learning,[1-3] developed this method in Britain in the 1940s, working at the Coal Board and later in hospitals, where he concluded that conventional instructional methods were largely ineffective. Revans suggested that people needed both to be aware of their lack of relevant knowledge and to be prepared to explore these areas with suitable questions and help from other people in similar positions.

Revans' theory of action learning was cited by Stuart Crainer as 'one of the 75 greatest management decisions ever made', in his book of the same title.[4]

Revans' action learning formula[1]

$$L = P + Q$$

where **L** is learning; **P** is programmed (traditional) knowledge; **Q** is questioning to create insight.

Q uses four 'major' questions:

- Where?
- Who?
- When?
- What?

and three 'minor' questions:

- Why?
- How many?
- How much?

Although **Q** is the cornerstone of the method, the more relaxed formulation has enabled action learning to become widely accepted throught the world. See Revans' book for international examples.

15.3.1 Action learning in practice

What is it?

Action learning is a process for bringing together a group of people with varied levels of skills and experience to analyze an actual work problem and develop an action plan. The group continues to meet as actions are implemented, learning from the implementation and making mid-course corrections. In other words, action learning is a form of *learning by doing.*

When should you use it?

Action learning may be implemented:

- To address problems and issues that are complex and not easily resolved.
- To find solutions to underlying root causes of problems.
- To determine a new strategic direction or to maximise new opportunities.

How should you use it?

- Clarify the objective of the action learning group.
- Convene a cross-section of people with a complimentary mix of skills and expertise to participate in the action learning group.
- Hold initial meetings to analyse the issues and identify actions for resolving them.
- Return the group to the workplace to take action.
- Use subgroups to work on specific aspects of the problem if necessary.
- After a period of time, reconvene the group to discuss progress, lessons learned, and next steps.
- Repeat the cycle of action and learning until the problem is resolved or new directions are determined.
- Document the learning process for future reference. Record lessons learned after each phase of learning.

Case study: London Borough HR managers

We worked with a London borough that used action learning as part of a two-year development plan for HR managers. The borough wanted their HR staff to practice working with real business problems as a basis for learning. Senior managers identified critical concerns meaningful to the borough and acted as sponsors for the action learning teams. Our task was to approach other boroughs and see if their HR personnel would be willing to participate and share their knowledge and experiences. Three teams met over a six-month period using the action learning methodology. At the end of that time, teams met with their sponsors and reported their results. Senior managers were delighted by the creative work accomplished by the teams and followed through on many of their suggestions.

15.4   Having a mentor

As you progress within your organisation it is prudent to have a mentor to call on for advice and guidance. In any organisation there is a great deal of hidden agenda, for example:

- Who really holds the power at a more senior level?
- What should you do if your immediate manager acts as a gatekeeper to your progression?
- If you have a good business idea that you believe is viable, and you can prove your business case but find the person you report into blocks all new entrepreneurial ideas, what should you do?
- When do they envisage, from their own experience, that new budget/research funding will be in place for x/y/z?

These are probably the types of conversation most of you want to have in your day-to-day working activities, but feel you can't necessarily take them to your manager (particularly if the reason you need to seek advice is because of their action, or more accurately, inaction!).

The ideal mentor is sufficiently senior to have knowledge, experience and clout. They need to feel that the process appeals to them and they need to have time to put into the process. The time need not be great. The key thing is their willingness to spend some time regularly helping you. Usually, if such a relationship lasts, it will start out one way – they help you – but may well become more two-way over the years; perhaps the person on the mentor side makes the decision to help partly on the basis of this anticipated possibility. A mentor is usually not your line manager.

Your mentor is there in an advisory capacity, to act as a sounding board and help you steer through the corporate jungle. After the first few years of a career, there is no reason why you cannot have relationships of this nature with a number of regular contacts.

In many blue chip organisations and also increasingly in the government sector, mentorship is a regular part of ongoing development. You may be allocated a mentor or able to request one. Equally you may need to act to create a good mentoring relationship. You can suggest it to your manager, or even take the initiative by directly approaching someone you think might be willing (or persuaded) to undertake the role.

Mentoring should be regarded as very different from, and very much more than, networking. The nature and depth of the interaction and the time and regularity of it is much more extensive. This is not primarily a career assistance process in the sense of someone who will give you a leg up the organisation through recommendation or lobbying, though this can of course occur. It is more important in helping develop the range and depth of your competences and improving your job performance, with this in turn acting to boost your career. A senior or well-connected mentor may also act as an early warning system when trouble is brewing.[5]

The difference between mentoring and coaching:

Form	Description	When to use it?
Manager-report coaching	Line manager spends time with direct report to work on some aspect of his/her performance	When the issue is a skill or aspect of knowledge where the manager can help. Should not be used in a punitive or other negative way
Mentoring	Enables an individual to follow in the path of a more senior colleague who can pass on knowledge and experience and open doors to otherwise out-of-reach opportunities	When senior managers have the time, experience and expertise to share with junior managers

15.5   Career planning

The role of a team leader is interesting, challenging and rewarding. In conversations with clients it has become clear that members of staff will often stay in the team leader role for a transitional period of time – say 18 months to two years, and then it will be time to figure out how to move on (Figure 15.1).

You need to take a proactive approach and plan your career.

Career planning is about self-assessment. Look within yourself to discover your interests, skills, personality traits and values. Also ask friends, family members or mentors if they identify the same qualities as you do. Simply ask yourself:

- What do I like to do?
- What activities do I find fun, motivating, interesting and enjoyable?
- What skills and abilities do I have or want to develop?
- What personal style or characteristics do I have that are important to me in the workplace?
- What purpose or goal do I want to accomplish in my career?

Figure 15.1   The wheel of career planning

15.5.1   Career exploration

You need to think through and investigate all the career choices, options, and opportunities available to you. Attend career fairs and develop your internal and external network. Talk to people in various careers; shadow or spend time with people in careers that interest you. Ask others:

- How did you move into this role?
- What is a typical day like?
- What type of training or education is required?
- What are the starting and average salaries?

Next, set some goals. Research careers that interest you to determine how to prepare for them and if you need additional or specialised training to be successful.

15.5.2   Job search

John Lennon's line, 'Life is what happens while you are making other plans' encapsulates a painful truth. There is perhaps no worse situation than to look back, saying to ourselves, 'if only . . .'.

Career planning is an ongoing process. Regardless of your age, it is important to assess where you are if you are to meet your goals and turn your dreams into reality.

Few people stay in one job or on one career path throughout their lives, and you may find yourself completing the process more than once along the way.

Active careerists rely not on good luck, though they will take advantage of any that occurs (their planning and positive attitude to the process ensuring that they can do so). Most often successful careers do not just happen; they are made. Recognise this, take action and see career success as something you create.

15.5.3   Caution! Perception is reality

Consider the following. You are asked to team lead an important and complex project. You as team leader have all the required attributes and necessary characteristics. Now you must present to the project stakeholders – board members and external customers.

If you come across as nervous or unsure, what might happen? One bad move or presentation may ruin or stall your career. Some say, 'never mind, it was a sound plan'. It's still not enough. Some may question the plan itself. Some may dismiss it. And what happens with the next interesting project? You are never considered, with obvious impact on your career.

This makes an important point. Many skills are essentially *career skills* – important to doing an effective job, and also to how people are seen and how they progress.

15.5.4   Active career management

Career skills that should be regarded in this way are:

- all aspects of managing people (remember your success comes from them)
- presentation and business writing
- influencing upwards
- numeracy and computer skills
- more general skills such as good time management

If you want to move up to a more senior role, you need to be able to indicate in any job application that you have the ability to take on the challenges of a more senior position. Below are five sample categories of challenge which may help you to position your application appropriately:

Challenge	Examples	Benefits
Start-ups	Heading up something new	Learning how to organise and get things done. How to select, train and motivate others.
Fix-its and turnarounds	Taking over a poor-performing business unit	Getting others to commit to change, and effectively orchestrating change. How to set up structure and control systems needed to turn business units around.

Projects and task forces	Troubleshooting Making a board presentation	Persuading, influencing and handling conflict. Learning what's important about unfamiliar areas and learning how to get things done through others.
Changes in scope and scale	Promotion with much greater responsibility	Shifting focus from doing things well to seeing that things are done well. Leading through persuasion and shouldering full responsibility.
Secondments	Moving to a role outside current experience	Understanding corporate strategies, culture and the broader context in which decisions are made. Thinking more strategically and less tactically

Self evaluation against these indicators

Competence	Role model behaviours – you demonstrate customer focus when you …	Possible development need – you fail to demonstrate customer focus when you …
We value our customers above everything else and aspire to make their lives richer, more fulfilled and more connected	– act as an ambassador for our brand values – create and promote a customer focused culture; keep customer satisfaction and convenience at the forefront of the team's thinking – innovate and pioneer new ways of serving customers	– allow your team to be/become introspective, bureaucratic or political, rather than customer-facing – take short-term financial decisions that penalise the customer – fail to recognise and reward customer service champions in your team
We must always listen and respond to each of our customers	– trust our customers, listen to them, seek to understand them and act to ensure that every aspect of our service to them is second to none – take action to recover from customer service failures to build or regain trust	– show a lack of understanding for customer needs, rarely talking to them directly or to the employees who deal with them daily – insufficiently prioritise corrective action when customers have problems

Figure 15.2   Example of competence application in a well known telecoms organisation

Barbara Hawker is Managing Director of Hurst Associates (Europe), an organisation that specialises in career planning and development. She shared with us her thoughts on career planning:

Case study: Careeer planning

If you consider the proportion of their lives that many spend at work, it is surprising how little thought, planning or action many people put into their careers and working life. While it is true that success is sometimes achieved by being in the right place at the right time; it would be good to feel that occasionally you had been instrumental in, or at least had some influence on being in that right place!

The days are long gone when you got promotion by waiting until it was your turn to move up within an organisation, irrespective of merit. As with most things, if you want to get on in your working life, it's up to you to make the effort.

And happily, many organisations are giving their staff the opportunity to do just this – by having formal appraisal systems in place during which ambitions can be aired, offering structured learning frameworks and encouraging you to attain recognised vocational qualifications associated with your profession.

But what happens when you feel 'stuck' and the organisation is being less proactive than you would like? Stuck or not, the answer to successful career planning is the same in both situations – take stock, make plans and, as a result, take control.

Understanding what you've got to offer is a good starting point for getting the job or career that could give you most satisfaction. Identifying which employers use those skills gives you the target organisations that you need to monitor or approach. And finally, making that important first contact with them or being prepared to respond to advertised vacancies may give you the opportunity to move and progress.

Many people have such full lives – with work, friends and family commitments – that they do not put themselves first on a regular basis. To work towards being more satisfied with your working life you need to take some 'time out' periodically and to spend that time carefully working through a number of issues.

Do you really understand what skills you have? Many people don't recognise their talents or think about their transferability to other roles. Forget about *what* you do – think about what *skills* you use undertaking your regular tasks. (Ask others for their views if you find this difficult to do by yourself; colleagues can usually tell us the good and the bad about our work and this can be really useful.)

Picture an ideal world, with no financial limitations, etc. Picture the type of work you would really like to do. What building bricks can you identify, e.g. size of organisation, type of environment and industry sector, the skills and experience you would use, the pace of work and role (team leader, specialist)? Can some of these be translated into the real world?

Can any of these hopes and wishes be achieved from your current role, with your existing employer? You may need to seek a training secondment to another part of the organisation to acquire the knowledge that would assist your internal progression. You may need to take further external training by classes, Learning or personal study. Use these activities as stepping stones to your ultimate goal.

If you need to move employment to achieve the next step, undertake research into those employers who appeal to you and who use the skills and enthusiasm you can demonstrate. How do they recruit? What agencies or processes do they use?

When internal or external opportunities to compete arise, do everything you can to make yourself the most suitable candidate. Few people really put themselves out when applying for jobs. They make half-hearted and scruffy applications; many do little preparation for interview and appear almost resigned to failure from the start. Think and act positively, show enthusiasm, care and interest, and above all take the process seriously. The amount of effort you put in will be mirrored by the level of success you experience.

Don't think you can achieve your ultimate goal in one step. Reality has to be overlaid on your aspirations. Aim for a number of incremental successes that are achievable and will move you along the right track. And keep at it. Win a new role and after a year review the situation and plan the next move.

Be realistic. Aim high but not too high. Raise your profile in the organisation. Your immediate boss might know you, but does their boss even know you exist or what you do? Particularly if you need to look for a sideways move within an organisation, get to know the key influencers who may be in a situation to 'talent spot' you.

Life, of course, doesn't always go to plan. The economic climate sometimes affects the number of vacancies available and personal constraints sometimes limit the amount of freedom we have to do something really 'outside the box' – but having a plan at least gives you a sense of direction, even if the timescale might have to be adjusted. Life sometimes throws a spanner in the works. Internal re-organisations and company mergers cause disruption, but they can also present unexpected opportunities to try something different. Make the most of what's on offer.

This may all seem very clinical. But remember, it's your career. Other people have their own careers to look after and don't have any responsibility for yours. If you want to be noticed for all the right things, volunteer for projects which will use your latent skills. Surprise those around you by being capable of things that they might not expect. And if this sort of opportunity doesn't arise, just make sure that your bosses recognise and understand your current contribution. Tell them about your achievements, any particularly difficult or unusual situations that you have handled well, or ideas you have to improve or develop current practices. It makes them sit up.

15.6   Chapter summary

- Career planning involves being a good networker, and enhancing your profile in the organisation by joining committees, or Action Learning sets.
- Have a mentor who can help you politically navigating your way through the organisation.
- Do a self assessment exercise to be clear about your own needs and requirements, and be pro-active in your search for a new role / career development opportunities.
- It is not always possible to make the career move that you would want in one go, so aim for a number of incremental moves that will steer you on the right track.

We hope you have found this handbook a useful and robust aid in your role as a team leader, and as a final resume, below are the key learning points to bear in mind in your day to day activities.

15.7   Final summary

- Develop a style of leadership that is appropriate for each particular situation.
- Make sure that the values, culture and environment are supportive to team development.
- Your recruitment policy and the way you induct new team members can make a substantial difference to the overall success of the team.
- Managing performance of team members requires feedback, a structured approach to the appraisal process, and appropriate recognition, reward and development.
- Problem solving should be part of a continuous improvement activity, not undertaken in a 'fire fighting' way.
- Delegate activities to members of your team, being aware of their capabilities, skills sets and competencies to undertake the task involved.
- Work at breaking down interpersonal and interdepartmental barriers through developing your influencing skills.
- Be proactive with regard to planning your own career, to create the work–life balance that is right for you.

'The only place where success comes before work is in the dictionary.'

Vidal Sassoon (1928–), British hair stylist

References

Preface

1. Hersey P. and Blanchard K. *Management Organisational Behaviour: Utilising Human Resources.* 3rd edn. New Jersey: Prentice Hall International; 1977.

Chapter 1

1. Crainer S. and Dearlove D. *The Ultimate Business Guru Guide: The Greatest Thinkers Who Made Management.* 2nd edn. Capstone; 2002.
2. Katzenbach J.R. and Smiths D.K. *The Wisdom of Teams.* Harper Business Essentials; 2003.
3. Robbins H. and Finley M. *Why Teams Don't Work.* Peterson's; 1995.
4. Mayo E. *Hawthorne and the Western Electric Company, The Social Problems of an Industrial Civilisation.* Routledge; 1949.
5. Dyer J.L. 'Team research and team training: a state-of-the-art review' in Muckler F.A. (ed.). *Human Factors Review.* Santa Monica, California: Human Factors Society; 1984. pp. 285–319.
6. Benne K. and Sheats, P. 'Functional roles of group members'. *Journal of Social Issues.* 1948;4(2):41–49.
7. Krech D. and Crutchfield R. S. *Theory and Problems of Social Psychology.* McGraw-Hill; 1948.
8. Margerison C.J. and McCann D. *How to Lead a Winning Team.* MCB University Press; 1985.
9. Colenso M. *High Performing Teams in Brief.* London: Butterworth-Heinemann; 1997.
10. Belbin R.M. *Management Teams.* New York: Wiley; 1981.
11. Tuckman B. 'Developmental sequence in small groups'. *Psychological Bulletin.* 1965;63(6):384–99.
12. Yalom I.D. *The Theory and Practice of Group Psychotherapy.* New York: Basic Books; 1970.
13. Senge P. *The Fifth Discipline: The Art and Proactive of the Learning Organisation.* London: Century Business; 1990.
14. Micklethwait J. and Woolridge A. *The Witch Doctors: Making Sense of the Management Gurus.* London: Mandarin Press; 1997.

Chapter 2

1. Blanchard K. with Johnson S. *The One Minute Manager: The Quickest Way to Increase Your Own Prosperity.* William Morrow & Co.; 1982.

Chapter 3

1. Carlzon J. *Moment of Truth*. Harper Perennial; 1987.

Chapter 4

1. Handy C. *Understanding Organizations*. Penguin; 1976.
2. Deming W.E. *Out of The Crisis*, Chapter 3. MIT Press; 1986.
3. Senge P. quoted in Napuk K. 'Live and learn'. *Scottish Business Insider*. January 1994.
4. Welch J. with Byrne, J.A. *Straight from the Gut*. Business Plus; 2001.
5. Masaaki I. *Kaizen: The Key to Japan's Competitive Success*. McGraw-Hill/Irwin; 1986.
6. Brandt D. 'L'Oreal factories play it safe'. *Industrial Engineer*. July 2007.
7. *Sunday Times*. 11 November 2007.
8. Kaplan G.S. and Patterson S.H. 'Seeking PERFECTION in healthcare: a case study in adopting Toyota Production System methods'. *Healthcare Executive*. 2008;23(3):16–21.

Chapter 5

1. Forsyth P. and Hailstone P. *Hook Your Audience*. P82. Pocketbooks; 2000.
2. Muir K. 'The Dark Ages'. *Sunday Times*. Supplement; 31 May 2008.
3. McCormack M.H. *What They Don't Teach You At Harvard Business School: Notes From A Street-Smart Executive*. New York: Bantam Books; 1984.
4. Peters T. *The Pursuit of Wow!* Vintage; 1994.
5. Biafore B. *On Time! In Track! In Target!* Redmond, Washington: Microsoft Press; 2006.

Chapter 6

1. Martin K. *The Global War for Talent: Getting What You Want Won't Be Easy*. Boston, Massachusetts: Aberdeen Group; 2007.
2. 'Recruitment Guide: Online Testing – Keep the Human Touch'. *Human Resources*. Haymarket Publications; October 2007.
3. 'Health & wellbeing: case study – The Consensus Organisation'. *Employee Benefits*. Centaur Publishing; 12 November 2007.
4. 'Case study: GSK outlines perks for staff'. *Employee Benefits*. Centaur Publishing; 8 October 2007.

Chapter 7

1. Maslow A.H. 'A Theory of human motivation'. *Psychological Review*. 1943;50(4):370–396.

Chapter 8

1. Younger J., Smallwood N., Ulrich D. 'Developing your organisation's brand as a talent developer'. *Human Resource Planning*. 1 April 2007;30(2).

Chapter 9

1. Learning & Development Staff Opinion Survey. International Survey Research; 2001.
2. Herzberg F. 'One more time: how do you motivate employees?'. *Harvard Business Review.* 1968;46(1):53–62. Also Reprint 87507.
3. Leitch S. *Leitch Review of Skills.* UK Government Publications; 2006.
4. Younger J., Smallwood N., Ulrich D. 'Developing your organisation's brand as a talent developer'. *Human Resource Planning.* 1 April 2007;30(2).

Chapter 10

1. Fuller E. *Customer Care, A Directors Guide.* Institute of Directors & Unisys Ltd; 1994.
2. Geller E.S. 'People-based leadership'. *Professional Safety.* 1 March 2008.
3. Tannenbaum R. and Schmidt W. 'How to Choose a Leadership Pattern'. *Harvard Business Review.* May 1973.
4. Belbin R.M. *Management Teams. Why they Succeed or Fail.* London: Heinemann; 1981.
5. Belbin R.M. *Team Roles at Work.* London: Butterworth-Heinemann; 1993.
6. Henry S.M. and Stevens K.T. 'Using Belbin's leadership role to improve team effectiveness: An empirical investigation'. *Journal of Systems and Software.* 1999;44(3):241–250.
7. Kanter R.M. *The Change Masters.* Simon & Schuster; 1983.

Chapter 11

1. Ohmae K. *The Mind of the Strategist.* McGraw-Hill; 1982.
2. Drucker P. *Innovation and Entrepreneurship.* Harper Collins; 1985.
3. Wellington P. *Kaizen Strategies for Customer Care.* Financial Times/Prentice Hall; 1995.
4. Juran J.M. *Quality Control Handbook* (expanded on the Pareto Principle). McGraw-Hill; 1951.
5. de Bono E. *Six Thinking Hats.* Little, Brown and Company; 1985.
6. Breen B. 'The six myths of creativity'. *Fast Company.* December 2004;89:75.
7. Kerzner H. *Project Management: Workbook.* John Wiley & Sons; 2003.

Chapter 12

1. Marriott S. 'Defogging your communication'. Kaizen Training Limited. LeadingMinds Tip, November 2006.
2. Berne E. *Transactional Analysis in Psychotherapy: A System.* Castle Books; 1961.
3. Bennis W. *On Becoming a Leader.* Perseus Books; 1989.
4. Covey S.R. *The 7 Habits of Highly Effective People.* Franklin Covey; 2004.
5. Levi-Strauss C. *The Raw and the Cooked.* Harper & Row; 1969.

Chapter 13

1. Clark L. 'Swatches: keep the brand new'. *The Statesman* (India). 16 March 2008.
2. Atkinson W. 'Novellus Realises Benefits of Early Supplier Involvement'. *Purchasing.* Reed Business Information; 10 April 2008.

3. Sinoo J., Willenborg J., Rozendaal S. *The Changing Face of Management*. Amsterdam Council of Organization Consultancy; 1988.
4. Robbins A. *Awaken the Giant Within*. Simon & Schuster; 1992.

Chapter 14

1. Tuckman B. 'Developmental sequence in small groups'. *Psychological Bulletin*. 1965;63(6):384–99.

Chapter 15

1. Revans R.W. *Action Learning: New Techniques for Management*. London: Blond & Briggs; 1980.
2. Revans R.W. *The Origin and Growth of Action Learning*. Brickley: Chartwell-Bratt; 1982.
3. Revans R.W. *ABC of Action Learning*. London: Lemos and Crane; 1998.
4. Crainer S. *The 75 Greatest Management Decisions Ever Made*. New York: AMACOM Publishing; 1999.
5. Forsyth P. *Kickstart Your Corporate Survival*. London: John Wiley; 2003.

Further reading

Colenso M. *Kaizen Strategies for Improving Team Performance*. Financial Times/Prentice Hall; 2000.

Ezrati M. *Kawai How Japan's Economic and Cultural Transformation will Alter the Balance of Power among Nations*. Aurum Press; 1999.

Goleman D. *Working with Emotional Intelligence*. Bantam; 1998.

Katzenbach J. *Teams on the Top*. Harvard Business School Press; 1998.

Lipnack J. and Stamps J. *Virtual Teams: People Working Across Boundaries with Technology*. 2nd edn. John Wiley & Sons, Inc.; 2000.

McGill I. and Beaty L. *Action Learning*. Kogan Page; 1992 & 1995.

Nemeth C. and Nemeth-Brown B. 'Better than Individuals' in Paulus P.B. and Nijstad B.A. (eds.), *Group Creativity: Innovation through Collaboration*. New York: Oxford University Press, Inc.; 2003. pp. 63–84.

Ocker R. 'Influences on creativity in asynchronous virtual teams: A qualitative analysis of experimental teams'. *IEEE Transactions on Professional Communication*. 2005;48(1):22–39.

Piszczalski M. 'Teamwork: Tools, techniques and troubles'. *Automotive Manufacturing and Production*. 2000;112(2):16–18.

Robins H. and Finley M. *Why Teams Don't Work*. London: Texere Publishing; 2000.

Wellington P. *Kaizen Strategies for Customer Care*. Financial Times/Prentice Hall; 1995.

Index